MORESHET

Journal of the Study of the Holocaust and Antisemitism

Claims Conference

This publication has been supported by a grant from the Conference on Jewish Material Claims Against Germany

MORESHET

Journal of the Study of the Holocaust and Antisemitism

Moreshet, The Mordechai Anielevich Memorial
Holocaust Study and Research Center, Givat Haviva

Alfred P. Slaner Chair in Antisemitism and Racism,
Tel Aviv University

11 | 2014

Moreshet, The Mordechai Anielevich Memorial was established in 1961 by: **Yehuda Bauer, the late Yisrael Gutman, Abba Kovner, Shalom Cholawsky, Ruzka Korczak, Akiva Nir, Yehuda Tubin, and Haika Grossman**

MORESHET
Journal for the Study of the Holocaust and Antisemitism

Published by
Moreshet, The Mordechai Anielevich Memorial
Holocaust Study and Research Center, Givat Haviva

The Alfred P. Slaner Chair in Antisemitism and Racism
Tel Aviv University

This publication has been supported by grants from

The Havatzelet Cultural and Educational Institutions

and from the Eta and Sass Somekh Family Foundation

 ISBN 978-1885881281

Editorial Offices and Administration
Givat Haviva, Doar Na Menashe 37850, ISRAEL
Tel: 972–4–6309248
E-mail: judymoreshet@gmail.com
The complete catalogue of Moreshet articles: www.moreshet.org

Design
Amy Erani www.multieducator.net

TABLE OF CONTENTS

BOOK REVIEWS

From the Editor

We are very pleased to present the English edition of *Moreshet, Journal for the Study of the Holocaust and Antisemitism,* No. 11 – April 2014. Articles previously published in Hebrew have now been translated and made available to readers of English. We hope you find them of interest. *The Moreshet Journal* is published together with the Alfred P. Slaner Chair in Antisemitism and Racism, Tel Aviv University, headed by Prof. Dina Porat.

The section on **Holocaust Research** begins with Dr. Rivka Elkin's article, "Hope abandoned: Jews left behind in Jewish Social Welfare Institutions in Nazi Germany." Elkin describes the efforts invested by the Jewish community in Germany under the Nazis to assist the aging Jewish population who were living under conditions of extreme distress. Elkin argues that the models of community assistance by Jewish welfare associations collapsed, "falling like ripe fruit into the hands of the Nazis, and were exploited by it to liquidate the population of the old age homes and loot their money."

Dov Schidorsky, has written a highly documented article on the Nazis' looting of Jewish Collections and Jewish Libraries in Germany. Supported by extensive documentation, the author describes how in 1939, the loot stolen from Jewish libraries was concentrated in Department VII of the RSHA. However, even before the establishment of the gold institution, the SS and SD security services had already despoiled Jewish collections of valuables, libraries, and archives.

Prof. Matityahu Mintz examines the attitude of the USSR to the Jews during the period between the Ribbentrop-Molotov Pact between

Nazi Germany and the Soviet Union in 1939 until Nazi Germany invaded the USSR in Operation Barbarossa in June 1941. He discusses one of the pacts signed during the second round of negotiations by Ribbentrop and Molotov in September 1939, called "The Protocol," signed by both parties on September 28, 1939, and examines the fate of the Poles and the Jews over the following two years.

Irena Cantorowich wrote a study on "Jews and Non-Jews in the Partisan Movement in Belarus, 1941-1944." She writes about how the movement was organized and about the Jews who joined its ranks, raising the problematic attitude of the partisans to the Jewish fighters who joined them and discusses the antisemitism among the partisans.

The section presenting **Research on Antisemitism** comprises two research articles: Graciela Ben-Dror on "'Traitors and Evildoers': Antisemitism in Spanish Textbooks during the Franco Regime in Spain since 1939," and Boris Morozov and Marina Niznik's article on "The Jewish Community and Electoral Campaigns in Russia, 2007-2008." Ben-Dror's article is based on a painstaking examination of dozens of elementary schoolbooks and high school textbooks from 1939 until the Vatican II council (1962-1965). The books show the extreme antisemitism of the Spanish education system, under the management of the Catholic Church as partner to the regime.

Morozov and Niznik have investigated the research methodology on the attitude of the Jewish community in Russia to the electoral campaigns, to the pro-Jewish or anti-Jewish declarations in the press during the elections, and the overt antisemitism during these campaigns.

The section devoted to **Reflections** contains a Personal Statement by Yehuda Bauer, noted Holocaust historian, on the state of research on the Holocaust. The **Book Reviews** features articles on several books published over the last few years which we feel are of special interest to our readers:

French Holocaust scholar Jean-Charles Szurek has written about the important work by Dariusz Libionka and Laurence Weinbaum, to appear in English as *Heroes, Hucksters and Storytellers: The story of the Jewish Military Union.* It was published in Polish as *Bohaterowie, hochsztaplerzy, opisywacze. Wokol Zydowskiego Zwiazku Wojskowego,* awaiting official publication in English. Szurek emphasizes the unique contribution by the two scholars to the history of the Jewish Military Union, based on extraordinary efforts at finding documents in the Polish archives opened after the fall of the Communist regime. The book begins with a deconstruction of all of the erroneous material written up to now, based on hoaxes by Poles who deceived Israeli writers. The book continues with a reconstruction of history of the ŻZW, which the authors sum up by stating that there is no surviving archival material on which to reconstruct the heroic story of this organization.

Cipora Hurwitz completed her review in praise of Bella Gutterman's biography of Zivia Lubetkin, *Zivia the One,* shortly before she passed away on Kibbutz Gal-On in 2012. She notes Gutterman's contribution to the research on Zivia as well as the importance of her historical context as described in the book.

Nili Keren reviews Yehuda Bauer's book, *Death of the Shtetl,* on the Jews of Eastern Europe before and during the Shoah, emphasizing the unique contribution of the research in supplementing missing information on the shtetls in eastern Poland, the lives of the Jews, their behavior and their destruction.

Guy Miron reviews the Alvin H. Rosenfeld book about *The End of the Holocaust*, which deals with how its remembrance will be shaped in the future, and the manipulations that have been carried out in order to preserve this memory. Guy Miron presents a very interesting review of the main questions and challenges that the book deals with.

We hope you find the articles contained in this issue informative and interesting.

— *Graciela Ben Dror, Editor*

Rivka Elkin

Hope Abandoned: Those left behind in Jewish welfare institutions in Nazi Gemany[1]

The fear that German Jewry would become one big old age home, in which a large and increasing number of the elderly and unemployed would be supported by the Jews of the Diaspora was raised among German Jewish communities in other countries as early as the beginning of the 1930s.[2] This was a prophecy soon to be fulfilled.

The aging process of the Jewish population of Germany accelerated with the panic to leave the country which was triggered by Kristallnacht. Three years later, in October 1941, when the deportations from Gemany began, the Jewish population was reduced to approximately 160,000. Out of these, there were about 50,000 senior citizens, many of them impoverished. The demographic picture of Germany Jewry was an inverted pyramid, with the 60+ population three times greater than the 0-19 age group.[3]

Historical research on emigration from Germany has up to now concentrated on those who succeeded in leaving - mainly the young and middle-aged. In contrast, the aging sector of the Jewish population has hitherto not received special attention from scholars.[4]

This article seeks to examine the fate of those left behind, and how the Jewish welfare institutions in the *Reichsvereinigung der Juden in*

Deutschland (the RJD, the umbrella organization of Jewish organizations in the Reich) addressed the increasing distress of the elderly until they were finally deported from Germany by the Nazi regime.

I argue in this article that the panic-stricken emigration which began after Kristallnacht could not, by its very nature, include the majority of the elderly; the outcome was that the remaining Jewish community was forced to address the problems of an aging population, most of whom were in need. Unfortunately, the assistance extended to the elderly who were left behind did not result in their survival, since the Nazis' policy of extermination, which worked to implement deportations using tactics of deception and dissimulation, stymied the efforts of the Jewish community to aid its elderly.

Emigration and the Elderly

The forced emigration during 1938 turned into full-fledged flight after Kristallnacht. When does emigration become flight? In her book, *Escape Before the Holocaust*, Ruth Zariz notes three characteristics:

1. Emigration to foreign, unfamiliar locations with different living conditions and climate.
2. The willingness to say goodbye to loved ones and splitting up families.
3. Employing unusual and illegal ways to emigrate.[5]

All of these markers took place in the Jewish emigration from Germany at the time. The places most available for emigration were in unfamiliar locations such as South America and the Far East, each with a different climate and an unfamiliar culture. Families split up and dispersed throughout the world. Parents and children were separated. Parents pushed their children onto the *Kindertransport*, taking them to countries of refuge outside of Germany, as the children hoped that their parents would join them in time. Thus the Jews remaining in Germany constituted the weakest sectors of the population. Children were a minority. Most were past "the golden age", found it difficult to obtain visas for other countries, and were unable to rejoin their families. Their age, health status and difficulty integrating into the workplace in their new countries worked against them. In time, many of them joined the circle of the needy, more in terms of needing nursing care than from the financial aspect. In many cases the elderly whose families succeeded in emigrating remained in Germany without family support, even if they had financial means. In these circumstances, they needed community support above all else.

In addition to the elderly were the children whose parents faced incarceration in concentration camps, and who were forced to leave Germany immediately with the hope that their children would be able to join them later. Others who remained were young people and adults who, for various reasons, were unable to emigrate, either because they did not have sufficient funds or the contacts which would enable them to obtain visas. Finally, among those left behind were those who feared to emigrate to unknown destinations or were in principle opposed to emigration, such as Julius Seligson of the Jewish leadership, who was against "emigration at all cost." In the summer of 1939, he stated, "Our opinion is that it is more honorable for a Jew to die a martyr's death in central Europe than to die in Shanghai. The first is a 'sanctification of God's name,' while the second is the result of a false emigration policy."[6]

The flight from Germany was a direct result of the persecution of Jews becoming more severe during 1938, culminating in Kristallnacht. Through decrees, laws and administrative ordinances, the isolation of the Jewish population from German society in public, social and economic life was accomplished, attacking the very foundations of its existence. Reducing possibilities open to Jews was intended to accelerate the impoverishment of Jewish society and intensify the pressure on Jews to leave the country. In other words, it was forced emigration. Indeed, many did seek ways to leave Germany. From 1939 on, the major Jewish institutions, predominant among them the RJD, which hitherto had supported methodical and well planned emigration, now concentrated its efforts in speeding up the emigration process. The path of emigration from Germany was strewn with obstacles: difficulties in obtaining financing (mainly the purchase of foreign currency), while after the war, destination countries had quotas, or entry visas conditional on work visas or guarantees posted by friends or relatives in addition to meeting all of the usual criteria such as health, age requirements and the like.

The war naturally increased the difficulties of traveling to other countries, and these difficulties became an obstacle mainly for the elderly among the population who wished to emigrate. The productivist trend dominating emigration during the early 1940s became a factor which made it difficult for those in their senior years to emigrate. There were programs for establishing settlements on a productive basis, primarily agricultural settlements, for thousands of Jews in foreign countries such as South America. In the Jewish newspaper it was argued that an immigrant who was not a candidate for a productive profession placed a burden on those who came after him and blocked the way for others.[7] An additional group harmed by this trend towards productivity were the disabled, including veterans of World War I, who, because of their injuries, were unable to become integrated into productive professions. A further difficulty was the problem in transferring the disability payments due to them to their new destinations.[8]

The community activities focused entirely on emigration and all its ramifications. Through the only newspaper the Nazi regime allowed to be published, *Der Jüdisches Nachrichtenblatt*[9] the Jewish establishment maintained contact with the community, disseminated the message of emigration, and publicized information vital to both those seeking to leave and to those left behind. The feeling of urgency to leave Germany was reflected in the focus on emigration until the onset of the deportations in October 1941. The newspaper expressed the desperate search for refuge as the gates of more and more countries were being closed to Jews. The call went out to the world, "Remove the barriers!" expressing the sense of emergency, to do everything to remove the obstacles impeding a speedy exit from Germany. The call went out to Jewish organizations worldwide, appealing to emotions and the tradition of Jewish solidarity in the face of danger, begging for assistance in money and actions to help get the remaining Jews out of Germany. During the first year of the war, as time was running short, long time immigrants who had already settled in various other countries were called upon to ask for family unification. Their mission was to purchase foreign currency and visas and to take those who had remained behind under their wing. Family unification was anchored in immigration laws in many countries, including the USA and Latin America, which, during the first period of the war, provided an opening for immigrants wishing to bring in family members, even if over the quota.[10] The newspaper's appeal to organize the emigration of guardians was intended to enable the absorption of immigrants who until now were denied entry to their target countries. Despite this, the problem of the family elders remained a problem.

The question, "When is old too old?" was discussed in one of the articles on the desire of the elderly to leave for overseas and strive to build a new life for themselves. Due to their frustration at the lack of sufficient concern for organizing their emigration, their complaints increased. The newspaper responded, not holding back from noting the complexity involved in "seniors" emigrating.[11]

One of the ways was to appeal to known or unknown sponsors living outside of Germany, but not through the assistance organizations, which in many cases tended to politely reject applications by the elderly. In a published feature, the newspaper provided useful ideas to the elderly: how to draw up an application for assistance appealing to a potential sponsor in the target country so as not to receive a refusal. From these proposals, we learn about the unique difficulties facing this population wishing to emigrate. Each application required a great deal of thought and caution. The writer must explain to the recipient why he presumes to write to him and why he expects his assistance. In like manner, it was important to explain to the sponsor that despite his age, the applicant would not become a financial burden.[12] These pieces of advice were not dissociated from reality, since outside of Germany, in the USA and in England, people still did not realize what was happening in Germany. Earlier immigrants from Germany did not fail to criticize new immigrants for leaving the country. Ruth Zariz found proof of this in a Letter to the Editor published in August 1940 in the *Aufbau*, the newspaper of Jewish immigrants in the USA. The letter complained:

> Their letters tear us up inside. We thought that they are being persecuted to death back there in Germany. We made feverish efforts to bring them over here over the needs of our families. Now we know that we should have left them in Europe. I am beginning to think that this streaming into here is unnecessary. If the reason is the loss of rights in Germany, after all, they should remain there and suffer for a few years until things become right again.[13]

Indeed, the efforts of many of the over 65-year-olds to leave Germany were unsuccessful. The increase in the number of institutions for this age group in relation to the increase in the number of child care institutions are evidence of this. In October 1941, just before the methodical deportations of the Jews of Germany, 231 social welfare institutions were active in the Old Reich: of them, 113 were old age homes and nursing institutions as compared with 23 homes for children and teenagers.[14] There is no doubt that we can attribute this reality to the demographic profile of the emigration.

During the 1930s, the ramifications of emigration on the age dispersion of the Jewish population was evident. The proportion of those age 65 and over doubled, increasing from 10.5 percent in 1933 to 23.3 percent in May 1939.[15] The greater the increase in emigration, the greater the pressure on the old age homes and nursing institutions, and the communities' efforts were diverted to solving the problem of those left behind.

In July 1939, the RJD, for whom social welfare was one of its functions, declared its commitment to act through its welfare institutions for the elderly, disabled, and welfare cases, to take care of them and provide a roof over their heads. Paul Epstein, a member of the RJD, described the situation:

> As a result of the accelerated emigration of young people, followed by the increased emigration of older people, the Jewish elderly remain without care. This fate also touches upon the disabled and social welfare cases. The Jewish welfare institutions are obligated to take care of this population and provide it with shelter. There is a great demand for institutions for the elderly and to those needing nursing assistance, so that the number of existing places is nowhere near what is needed, despite the density of housing in institutions we recently introduced, which made the intake of a larger number of applicants possible.[16]

The declarations by the communities and the RJD came face to face with a reality that made it very difficult for them to realize their promises. In 1939 there were no vacancies in the old age homes or in the social welfare institutions in the three largest Jewish communities – Berlin, Munich and Hamburg. The number of those on the waiting lists for moving into existing institutions, for example in Berlin, was twice as high as the occupancy in existing institutions, which were filled to capacity.[17]

This gap between the huge, increasing pressure to be accepted into old age homes and between those who found available places accompanied the Jewish welfare and nursing institutions throughout the entire period of the Nazi regime, from November 1933 until the deportation of this population.[18]

After the declaration of war on September 1, 1939, the Berlin community acted to find places in old age homes for those aged 65 and over. One option was to convert the children's homes, which were being emptied during this period, into old age homes, in light of the fact that many children were being sent out of the country since early 1939.[19] This proposal was removed from the agenda when the vacant childrens' homes were required for the intake of children from other communities that could no longer maintain educational institutions due to the small number of children remaining, and the lack of finances which made them impossible to rebuild.

Some communities, such as in Mannheim, took over the old age homes in the service of emigration. The district rabbi, Dr. Ernest Stekelmacher, asked the old age homes to take in the elderly relatives of those who were about to emigrate so that the latter could stop worrying about them and be heartened by the knowledge that their elderly relatives were not being abandoned to their fate but would be given care and shelter.[20] Another stance was represented by Moritz Henschel, chairman of the Berlin Jewish community, who was vehemently against the pressure applied by those leaving the country to place their relatives in old age homes before leaving. What they considered first and foremost was the housing pressure and paucity of vacancies in the old age home in Berlin. According to Henschel, a family is committed to their parents' fate, and should take the grandparents with them or have them join them later. He further argued that obviously, those who were unable to do so should make sure to have their parents admitted to an old age home before leaving the country, for their own peace of mind. Not every case was urgent, and some elderly parents were still capable of independent life and did not need an institutional environment. When the time came, should it be necessary, they would take care of them. Henschel emphasized that they should trust the administration. "Our concern for the old age homes is not only administrative, but a human obligation."[21]

Despite the limitations, from 1939-1940 the number of old age homes and nursing institutions increased. Thirty-two new institutions were established, the majority by the RJD. Some were located in communities outside of Berlin in vacated children's residences. In 1941 there was a decrease in the number of these institutions, arising mainly from the supervisory body's policy on emptying Jewish institutions and concentrating their populations in the existing institutions. The process of concentrating the elderly population in institutions continued until they were deported from Germany in the summer of 1942.

Concentrating the Elderly population in the social welfare institutions

There were two aspects to concentrating the elderly in the social welfare institutions: one was the nursing aspect, which was the responsibility of the Jewish social welfare department, while the second was government policy, which strove to concentrate all the Jews remaining in Germany until the "final solution" of eliminating them entirely.

Unlike Eastern Europe, there were no ghettos established in Germany. Despite this, a process of 'ghettoization' did take place, motivated by legislation and decrees that in practice forced the concentration of Jews and disinherited them, looting their assets. Thus, for example, the Tenant Protection Law, which from 1939 prohibited Jews from being protected tenants during their rental period,[22] and the "Aryanization" laws, which nullified Jewish ownership of property. As a result, the Jews were separated from the few assets still remaining to them. The Jewish community of Berlin found it difficult to cope with the eviction of Jews from buildings they owned, an action implemented as part of the construction plan for the city headed by architect Albert Speer, following the order that required maximum exploitation of the Jewish community's property.[23] Under these circumstances, the Jews were forced to concentrate most of their institutions and community services primarily in buildings belonging to the community.

Aryanization greatly harmed the Jewish welfare institutions, whose buildings were targeted by Nazi party officials and local government authorities.[24] Indeed, Goering's directive of December 1938 prohibited the Aryanization of the welfare institutions, fearing that those expelled from these institutions would become a public burden,[25] but in late 1940, the directive expired, giving the Nazis the opportunity to take them over.

The result was a sharp increase in the density of occupancy in the institutions which still had room. The old age homes were no longer welfare institutions for the aging population, but places of refuge to enable survival. During the transports which began in 1941, they ceased to serve their originally unique purpose. From then on, institutions of various types were all concentrated under one roof. Life was conducted in a constantly changing reality. With the issuance of an institution evacuation order, residents were redistributed into intake institutions whose ability to absorb new residents could not meet demand, while the caregiving staff was constantly being reduced. The result was the steady deterioration of the residents' standard of living. Residents of the institutions were asked to share their rooms with the new people joining them, to give up some of their furniture, and assist in the maintenance jobs of the building.

The difficult reality outside and inside the old age homes was expressed in the appeal by Hannah Karminsky, one of the heads of the Welfare Department of the RJD, to residents of old age homes in January, 1941. In her letter, she expressed the mutual relations between what was going on outside of the home and how matters were conducted inside the residential institution. She called for the reinforcing of social solidarity in the face of the ever worsening distress. She wrote:

> Most of you have already been living in our old age home for a long time. Perhaps you have sometimes thought, 'We were better off before this! We had private rooms, good food, we could rest, and our every request was fulfilled. But when we were forced to ask you [to share space], you crowded closer together. some of you shared your rooms with one or more, since you knew that these people had no other place to live, because if not, they would have to be satisfied with housing that we feel would be unbearable, while our situation is relatively much better … As you know, there are still many other people waiting for intake in an old age home. It is nearly impossible for us to establish new institutions, consequently we must ask you once again to increase the number of beds in the existing old age homes. Please, we beg of you to accept our request and make it easier for the institution's administration to implement this call. Please crowd closer together once again!

Later on, Mrs. Karminsky called upon the residents to participate in carrying out tasks necessitated by the new decrees that affected the manpower in the institutions:

> Conditions in the old age home will require you in future to increase your assistance in the housework. You are aware of the fact that on the outside each able-bodied person is being recruited into the 'labor service' – the age ceiling is now high. This is why in the future we must attempt to run our old age home through its own residents, and recruit younger people only for the hard work.[26]

Since the entire Jewish welfare system, both private and public, was now the responsibility of the RJD, it was incumbent upon the social welfare department to find alternate solutions for those evacuated from the institutions. One option was to raise the admittance age to 75 and over, others included establishing emergency residences – *Notheime* - in Berlin, turning the Jewish community center of Worms into an old age home and putting residents in Köln into huts.

As we said, the lack of buildings during the war was inverse to the rise in the number of residents needing shelter: in 1940 there were 122 institutions in the entire Reich housing 7,101 elderly residents. In the spring of 1942 the number of institutions went down to 111, but the number of residents rose to 8,360. This phenomenon was due to the continued confiscation of buildings for government needs, and because of the regime postponed the date of the deportations of the aged, ill, and disabled until the summer of 1942. At the same time, an order concentrated all the remaining elderly after their families were transported, so that they would disappear from public view and be within reach when the time came (primarily outside of Berlin).[27]

The directive created a paradoxical situation in which the need arose to establish new old age homes precisely during the deportations.[28] The situation in the Köln Jewish community is the one that most illustrates the implications of the directive. At the height of the transports in January 1942, six old age homes were established in Köln for a population of less than 6,000.[29] In April 1942, the number increased to ten, housing 1,139 people. These constituted about one half and perhaps more of the population remaining in the community. This proportion is a direct result of the pace of the transports over those three months.[30]

As the lack of buildings housing the needy grew greater, the distress of the population increased. They were not being deported, but remained behind. The index of density noted not only the average number of residents sheltered in one institution, but also illustrates the relationship between the occupancy capacity of the institution in the past and the number of residents crowded into it in the present period. Following are a number of examples for comparison between the number of residents in old age homes in various communities throughout the Old Reich in 1933 and their numbers in these institutions in April 1942:

In the abovementioned old age home, in Köln, in April 1942 392 people were crowded into the institution, while in 1933 only 74 people lived there. In an old age home in Würzberg, which in 1933 housed 50 elderly, the number rose to 360 residents after two institutions were merged.[31] In an old age home in Aachen, the number of residents increased from 35 to 117. In an institution in Frankfurt-am-Main, 131 men and women were crowded into a residential home occupied by 18 women in 1933. In Hamburg, there were 115 people living in the old age home, while in 1925 there were 47 residents.

The density index in Berlin reached 112 people in April-May 1942 on average per old age home, as compared to 67 in the Reich. This shows a greater pressure of density of occupancy in the city which was a regular destination for internal immigration since the Nazis rose to power. To lessen the burden on the old age home, healthy people were placed in private apartments in its vicinity, so that during the day they could stay in the institution and benefit from their services. For adults who could no longer maintain independent households but were not ready to live in an institution, foster family placement was suggested. For those who, due to illness, old age, or other reasons, needed assistance in their home, it was proposed that they have home care by teams of doctors and nurses sent by the welfare department. To make it easier on the nursing institutions who cared for TB patients after the sanatoria were closed down, joint residence was arranged in special apartments, with everyone involved also sick with TB. In general, there was no need to subsidize these residences, which were funded by those with means.[32] This idea later served as a model for old age homes and social welfare institutions.

In June 1942, approximately eight months after the beginning of the transports of the general Jewish population in Germany, the *Alterstransporte* – the transports of the elderly -began to Theresienstadt, nicknamed "ghetto of the elderly," and was defined as a location of "resettlement." Concentrating the elderly in the social welfare institutions prior to this date made it easier to transport them.

Postponing the date of transporting this sector of the non-productive population is surprising and raises many questions, especially when taking into consideration the fact that in Eastern Europe the non-productive members of the population were among the first to be transported.

The basis for this postponement may be found in Eichmann's instructions after the Wannsee Conference, at which it was decided who would not initially be "sent to the East". This category included the elderly from age 65 and older and nursing cases aged 55-65, defined as those unfit to be deported and whose "handling" was postponed. The expression "unfit for transport" requires an explanation, and might be included in the possible motives behind the postponement of the transport of the old and ill. We can note not only the fact that they were concentrated in institutions and immediately accessible, and the difficulties involved in moving them, but also the regime's strategy of deception, since deporting the elderly and ill as part of the transport of the general Jewish population of Germany could undermine the Jewish public's belief that this was merely "resettlement for work."

The policy of deceit that created the illusion

In a situation of uncertainty and disinformation, transport to Theriesenstadt instead of "the East" was considered a special privilege and a seeming reprieve from death. At the Berlin Jewish Hospital, they even held a party when hearing the news of the transports of patients and welfare cases to Theresienstadt.[33]

Transporting the elderly to Theresienstadt involved deceiving the people and looting their remaining money. Those being transported were forced to sign a "Purchase Contract" to ensure them a place in an old age home supposedly located at the new destination to maintain them in their old age.[34]

Resettlement of the elderly as defined by transport to Theresienstadt nurtured illusions and thus prehaps prevented the phenomenon of suicide among the elderly. Losing hope when forced out of the place they considered their homeland and fear of the unknown was undoubtedly one of the factors of suicide. These considerations were apparently repressed when faced with the opportunity of moving into an old age home in the new location. It is reasonable to assume that the hope factor explains why the elderly were not among the suicide statistics among the Jews of Germany. Sociologists point to advanced age and social conditions as characteristic of the suicide profile during that time. Among the factors listed are being severed from family due to emigration, social isolation and the emptiness called "social fear.'[35] There is a reasonable basis to assume that the framework of life in the old age home reinforced the feeling of security of the aging population, provided the elderly with a social group, care and protection, and so prevented suicide.

The idea of having people sign a "Purchase Contract" for a place in an old age home arose during 1940 in the welfare department of the RJD, and was necessitated by events. The goal was to cover the cost of the maintenance of the needy from the income of those with means who were signatories to the contract. The meaning of the contract was that, "those with means would support those without means."[36] In early 1941, there were still those among the old age home residents who could deposit a one-time sum of 2,000 Deutschmarks as an entry fee, or monthly payments of 150 marks, but alongside them were needy people who could not manage the required expenses. The proportion between those paying the entire sum and those depending on the support of community assistance was 30 percent as compared to approximately 20 percent of the overall number of old age home residents.[37] The heavy financial burden involved in administration of the old age homes and the desire to fairly and justly divide this burden gave birth to the idea of placing the obligation on those who were able to finance those institutional residents who lacked the means. Following the constant increase in the number of residents, the need to increase the total income arose, so that the old age homes could also take in those without means.

During 1941 the welfare department of the RJD made a decision to require a "Purchase Contract" to ensure a place in the old age home and a "Progressive Acceptance Agreement." From now on, the intake of those with means into the old age home was conditional on these terms. According to this decision, residents in institutions would henceforth pay what they could. The sum of 150 marks was calculated according to a monthly payment of 150 marks according to the life expectancy, in addition to the entry fee of 2,000 marks which was the fee up to then.[38]

An idea of this type had already come up during the 1930s but it was only during the 1940s, due to the force of circumstances, that it was implemented.[39] The circular sent to districts and communities noted that "Jewish welfare" would succeed only if all of the Jews with means still remaining in place would fulfill their obligations to society. Those with means would bear the burden of the expense of care for both themselves and their needy fellow residents. The deposits were administered by each institution in a special account for the institution, and were used for a general fund. The balance from the "Purchase Contract" was recorded in each resident's personal file. The original idea of signing a contract as a condition for intake in the old age home was based on the social principles of mutual aid and fair distribution of the burden of care.

On the surface, the community operated a fair and value-based mechanism; however, as time passed, it became clear that the process brought about the opposite of its intentions. The beginning was in April 1943, with the demand to sign a written agreement with no right of appeal.[40] Refusal to sign the contract was likely to result in "resettlement to the East" instead of Theresienstadt. In April, the deportations of the elderly to Theresienstadt had not yet begun (they began only that summer) and, as we said above, it was considered a preferential location, and therefore gave an opening to defraud the elderly out of their property without recompense.[41] Adjacent to the time of this directive, the RJD was required to report to the authorities on the extent of the "acquisition documents" signed, and on the amounts of money deposited.[42] In contrast to the previous procedure, those who did not have the necessary amount of money were also required to sign the contracts.[43]

About one month after the onset of the Eldertransports, the plan of seizing most of the capital in the possession of the signers began to be implemented.[44] The funds deposited by the signers included not only cash and securities, but also promissory notes, life insurance policies and the like.[45] From September 1942 on, it was decided to allow those contacting the organization to pay also with notes carrying liens.[46] The funds were concentrated in a special account labeled "H" and transferred to a bank in Prague, where they were seemingly registered in favor of the "Independent Jewish Administration in Theresienstadt," but in practice were controlled solely by the supervisory body, i.e. Eichmann's department at the Reich Main Security Office.

Up until October 1942, 36m marks had accumulated in the account, the amount having increased at a tremendous rate as the pace of the deportations stepped up that summer.[47] The new format of the contract served, among other functions, as legal cover for the RJD in case it moved signatories from one location to another, and also if there were changes in the conditions at the location. Avraham Barkai's research raised the hypothesis that the RJD preferred to cooperate with the supervisory body in this matter so as to retain control – albeit limited – over the capital, if only to protect it from falling into the hands of the Ministry of the Treasury.[48] On the other hand, concentrating the capital of the residents of old age homes apparently owned by RJD also served the supervisory body in attempting to bypass the Ministry of the Treasury as the legal beneficiary of the Jews deported from Germany (pursuant to Directive 11 of the Law of Citizenship of the Reich), and thus get hold of their cumulative capital. If we accept Barkai's argument, then in this case it created a situation of mutual interests between the supervisory body and the RJD, each party with its own motives.

The directive, which made it obligatory to sign the "Purchase Contract" for a place in the old age home in Theresienstadt, could serve as an example for an idea born in the Jewish Welfare Department that was cynically exploited by the supervisory body for its own needs. This meant carrying out orderly deportations undisturbed, and concentrating the capital owned by the elderly in a special account, which it controlled. The obligation to sign a contract, which they never intended to honor, demonstrates the dual nature of the Nazi regime, i.e., the superficial appearance of a country run by law, masking the commitment of illegal persecutions to promote the government's interests.[49] According to Saul Friedländer, this was "double talk": the external camouflage that in the future would aid in the success of the 'final solution.'"[50]

Those who signed the document were, however, then confronted by the reality of Theresienstadt, and the illusion dissipated very quickly. Dr. Hadda, director of the Jewish Hospital of Breslau, innocently thought that signing the purchase contract would provide the right to have at least basic humane conditions in the new settlement. Instead, signatories were thrust into a reality in which thousands of people were crowded into old barracks, where they slept on mattresses, at best.[51]

The shock of encountering the reality of Theresienstadt was excellently expressed by Julia Neuberg in a poem she wrote, not without cynical humor. She herself was murdered in Auschwitz at age 85. After describing the harsh and exhausting journey to the camp, she wrote:

> We searched for our old age home,
> it was already nearly midnight
> when we came along the road
> and then finally came to the right place.
> We would have settled ourselves in place very willingly,
> but where? There are no chairs, no benches,
> no bedside tables, no table, no cupboard.
> The rooms were entirely empty,
> as the disappointment increased.

> We settled our weary limbs
> to sleep on the ground…
> and then came the first lunch,
> the black potatoes, food fit for dogs,
> there's no thinking about any sort of comfort.[52]

In June 1942, the "Eldertransports" to Theresienstadt began and by January 1943, many of the 111 old age homes still existing in the Old Reich in April 1942 were liquidated. The last to be deported were the residents of the last three welfare institutions in Berlin.

Residents of the nursing home on the Auguststrasse were deported to Theresienstadt on May 29, 1943, together with the home's director, Rivka Oberlander, whose husband and two sons were already living in England and in Eretz Israel respectively.

The last remaining elderly were deported on June 16, 1943. These were the aged patients from the Jewish hospital, and residents of the Elsaesserstrasse nursing home. Simultaneously, the Jewish community of Berlin was destroyed and the last remaining employees deported.[53] The delay in the deportations gave both caregivers and patients the illusion that they would not be harmed. Thus, for example, the evidence given by Charlotte Holzer, whose mother (56-year-old Margarete Abraham) was among the nurses caring for the residents of the nursing home. She begged her mother to flee and go into hiding, but her mother did not believe that they would harm helpless patients, and was thus deported along with her charges.[54]

The feeling of frustration and helplessness of the Jewish community's caregivers as they faced their end was reflected in Hannah Karminsky's letter to Dr. Hans Schaeffer in Sweden, in June 1942:

> This work is no longer able to bring satisfaction. Already it has almost no connection whatsoever to what we

> have previously understood as social work. When we are dealing with people and not with properties or monies the evacuation is especially difficult. But since we are dealing with human beings, there are here and there moments that become meaningful by our very presence, which is what brings us satisfaction.[55]

And thus, the only ray of light in the bitter end of the residents of old age and nursing homes was the presence of the caregivers, who remained at the side of their patients.

From a quantum of solace to hope abandoned

With the liquidation of the old age homes and nursing institutions and the deportation of their residents, which was shrouded in deceit, the last hope of survival was robbed from the elderly even after they were resettled outside of Germany. With the liquidation of the institutions, the efforts that the Jewish community had invested in its concern for this population under conditions of deep distress were all in vain. Furthermore, it became clear that the sociological models of mutual aid and sacrifice for others implemented by the Jewish social welfare department in the welfare institutions in its struggle for their survival, even in the very narrow remaining room to manoeuver, fell like ripe fruit in the hands of the Nazi regime, who exploited them to realize its goal of liquidating the population of the old age institutions and looting their money.

The fate of the elderly remaining in Germany can be compared to people trapped in a whirlpool of persecution threatening to drown those caught in it. The only hope for survival was emigration, impossible for the elderly. The assistance provided by the Jewish social welfare institutions made it possible for those left behind to keep their heads above water, before the Nazi regime exploited the welfare system to use it to haul in all those trapped in the undertow and hurl them into the abyss.

Notes

1 The current article is based on my lecture at an evening conference on "70 years since the deportation of the Jews of Germany," held at the Leo Baeck Institute, Jerusalem, on November 3, 2011.

2 Finanzbericht d. RVt., 1936, The Central Archives for the History of the Jewish People (CAHJP) M14e, 163.

3 51,124 people age 60 and above as compared to 16,511 people aged 0-19.

4 For a more extensive discussion of Jewish social welfare under the yoke of the Nazi regime see Rivka Elkin, "The Heart Beats On: Continuity and change in the activities of Jewish social welfare organiztions of German Jewry under the Nazi regime, 1933-1945" (Jerusalem, 2004).

5 Ruth Zariz, *Escape Before the Holocaust,* (Tel Aviv, 1990), 40.

6 Yehoyakim Kochavi, "The last stage in the history of the Jews of Germany, 1938-1943," in A. Margaliot and Y. Kochavi (Eds.), *History of the Holocaust: Germany,* Vol. I, Part III (Jerusalem, 1998), 311-312.

7 "Chancen der Übersee-Wanderung," *Jüdischer Nachrichtenblatt Berlin (JN),* February23, 1940.

8 Ibid., p. 2; Dr. Oscar Götz, "Auswanderung der Körperbehinderten," JN Berlin, January 27, 1939.

9 Rivka Elkin, *Yalkut Moreshet* 88 (Sept. 2010), 59.

10 "Die Vewandeten – Nachwanderung," *JN,* Berlin, September 19, 1940, 1.

11 Ibid., 2.

12 "Briefsteller für älootere Auswanderer," *JN*, Berlin, September 19, 1941, 1.

13 Quoted by Ruth Zariz, 72.

14 Elkin, *Heart,* 268, 269.

15 Avraham Barkai, "Emigration and its ramifications," in M. Meier (ed.), *History of the Jews of Germany in the Modern Era, Vol. IV* (Jerusalem, 2005), 223.

16 Yad Vashem Archives (YVA) MF1 (1) No. 96, July 7, 1939.

17 Bundesarchiv Potsdam (BAP), 75C Re1, YVA MF 1 (1), No. 98, July 7, 1939.

18 Elkin, *Heart,* 126.

19 Margaliot and Kochavi, Vol. I, 310; see also Zariz, 139, 153, 154, 158; Elkin, *Heart,* 219, 220.

20 Ernst Steckelmacher, "Jüdischer Altersheime im Diesnste der Auswanderung," JN Berlin, September 6, 1940, 3.

21 Ibid., 3; Moritz Henschel, "Altersheime," JN Berlin, July 14, 1939.

22 "Gesetz ueber Mietverhaeltnisse mit Jüden," April 30, 1939, 1981, III, 190; Joseph Walk (Ed.), iDas Sonderrecht fuer die Juen im N.S.-Staat, (Heidelberg/ Karsruhe, 1981).

23 Wolf Gruner, "Die Reichshaupstadt und die Verfolgung der Berliner Juden 1933-1945," in Reinhard Rürop (ed.), *Jüdischer Geschichte in Berlin (*Hentrich, 1995), 246.

24 On the seizure of Jewish assets, see Elkin, *Heart,* 287-292.

25 Ministerialpraesident Generalfeldmarschal Goering, Beauftragte f. den Vierjahresplan, Geheim, December 28, 1938, YVA, 051 OSOBI.138,3. Vermerk, 051.) S)BI.109, February 23, 1939, 8.

26 "An die Insassen ünserer Altersheime! v. Karminski J., Abt' Fuersorge RV, January 1941, BAP 75C, Re1, No. 484, 157.

27 RV Personalverwaltun Bericht an RSHA May 8, 1942, BAP 75C Re1, No. 125, 50, #IV/B.

28 Ibid.

29 RV, Fuersorgeeinrichtungen bei denen die Saldierung der Einnahmen u. Ausgaben Stattfindet, veraenderungen, January 1942, BAP 75C Re1, No. 125, 128.

30 BAP 75C re1, YVA MF(480(4), 244-246.

31 BAP 75C re1, YVA MF 480(4), April 1, 1942, April 15, 1942, 244, 245, 246.

32 Prochnick Bericht, August 18, 1941, YVA 0.8/14, 26-30.

33 Bruno Blau, "Vierzehn Jahre Not und Shcrecken," September 12, 4..., YIVO, N.Y.B, 95.

34 Karl Eisemann, BST Baden-Pfalz Reichsvereiningung, an d. Mitarbeiter, August 16, 1942, Leo Baeck Institute, New York, AR, C-72412048, 2.

35 Konrad Kwiet, Helmut Eschwege, Selbstbehauptung und Widerstand, Desutsche Juden in Kampf um Existenz und Menschenwürde 1933-1945, 210, 213.

36 BAP 75C Re1, YVA MF 480(4), 244-246.

37 Prochnick Bericht, op.cit.

38 RV -Erleuterungen zuMuster des Heim-Einkauf-Vertrages, BAP 75C Re1, No. 484, 145, N.D.

39 Elkin, *Heart,* 134.

40 Reichsvereiningung, April 30, 1941, BAP 75C Re1, YVA MF 480(4), 249, RV May 26, 1941, 250.

41 Karl Eisemann, BST Baden-Pfalz Reichsvereiningung, an d. Mitarbeiter, August 16, 1942, LBI, NYC-72412038, 2.

42 Aktennotiz F 70, April 20, 1942, BAP 75C Re1.

43 Protokol RV May 11, 1942, BAP 75Cre!, YVA MF 2(1), 21.

44 Protokol RV July 16, 1942, BAP 75C Re1, YVA MF 3(2).

45 One example is "Für Konto "H" sind folgende Effekten eigegangen, August 14, 1942, BAP 75 Re!, No. 484, 37.

46 Protokol RV September 7, 1842, BAP 75C Re1, YVA MF 3(2.

47 Avraham Barkai, *Vom Boykott zur Entjudung* (Frankfurt a.M., 1988), 194, 196.

48 Ibid., 201.

49 Rürop, "Das Ende der Emanzipation: Die antijüdische Politik in Deutschland von der Machtergreifung bis zum Zweiten Weltkreig," in A. Pauker (ed.), *Die Jüden im NS Deutschland 1933-1943* (Tübingen, 1986), 104.

50 Saul Friedländer, *Nazi Germany and the Jews: The Years of Persecution 1933-1939* (Tel Aviv, 1997), 346 (Hebrew).

51 Hadda Sigmund, "erinnerungen eines Jüdischen Arztes, Mitteilung des Verbandes ehemaligen Breslauer in Israel," 59, 1995, 5.

52 "Mrs. Julia" (Neuberg), "Poem on Theresienstadt," YVA, 02/406, 3. English translation from the author's Hebrew translation of the German.

53 YVA, 08/145, 28.

54 Charlotte Holzer, "Bericht ueber Margarete Abraham," November 15, 1957, WL PIII C, No. 511.

55 "Letters from Berlin, 1942," *LB Year Book,* 1957, 312.

Dov Schidorsky

Jewish Libraries under German Occupation: The Reich Security Main Office as an Agent of Nazi Looting*

From the moment it rose to power, the Nazi regime in Germany was active in eliminating Jewish culture and eradicating its memory. It did so by looting, burning, and destroying Jewish spiritual treasures and by purging German libraries of their collections of Jewish literature, dispersing their contents throughout the ghettos and concentration camps. At the same time, the Nazi regime also labored to assemble and safeguard complete unique Jewish collections, which they regarded as a foundation for future research that would bolster the legitimacy of the extermination of the Jewish People and its heritage. The agents of looting of Jewish books included, among others, the civil administration of the *Generalgouvernement*, the Künsberg Special Commando, Julius Streicher and the editorial staff of *Der Stürmer* (the newspaper he founded and published), the Central Office for Jewish Emigration, the Book Clearing Office (Bücherverwertungsstelle) in Vienna, and special units of the S.S. However, the two primary agents of looting were the institutions, the organizations and the military units controlled by Alfred Rosenberg, and the Reich Security Main Office (*Reichssicherheitshauptamt*), or RSHA.[1] Within this agency, the Nazis set up the central library of these looted collections, which was intended to serve all the departments of the RSHA. Other specialized libraries such as a legal library, a research library for Eastern Affairs, and the SD school library were also established.

This article focuses on the goals, duties, and activities of the RSHA library, as well as the RSHA personnel and SS officers who were directly involved in shaping the policy of looting, carrying out the looting itself, and in making decisions regarding the fate of the looted objects. Library workers included a group of Jewish forced laborers who operated under SS supervision in closed concentration camp conditions. The confiscated collections provided the basis for Nazi research of Judaism. Research products included lists of individuals wanted by the Gestapo, reports on the confiscation operations, and manuals for the looting operations. These and other documents were placed at the service of the regime and supported the process of legitimization of the extermination of the Jews.

This article is based on archival documents and on secondary research literature. One particularly important source is the testimony of Ernst Grumach, who was appointed to head the group of Jewish forced laborers assigned to the library. Grumach's testimony provides an account of the group's work and addresses the attributes and personalities of the SS officers who supervised the forced laborers.[2] It also documents the atmosphere of fear and apprehension in which the forced laborers worked, as well as Grumach's ultimately unsuccessful efforts to save them from deportation to the East. The article also makes use of the following documents: The first is a report by Karl Burmester, who served as the director of the RSHA library from 1942 to 1945,[3] and the second is a report resulting from the interrogation of Paul Dittel by British intelligence. Dittel was deputy director, and, from 1943, director of Department VII of the RSHA.[4]

I. A Central Library for the Confiscated Collections

In early 1939, the looted collections of the Jewish libraries were consolidated in Department VII of the RSHA. However, even prior to the establishment of this entity, the SD–the intelligence agency of the SS and the Nazi party in Germany had undertaken mass seizures of collections, libraries and archives that were not necessarily owned by Jews. The units of the RSHA, and the entities that preceded them in this task, underwent numerous changes that continued until the end of World War II. To understand the background against which these changes took place and the role of the library of the looted collections within the RSHA, some milestones in the evolution of the library and its organizational structure are discussed.

Department VII of the RSHA had its roots in the secret intelligence service of the SS– a service that was already in existence during the period of the Weimar Republic. At the time, the Nazi Party was subject to attacks by its political opponents, and its members therefore engaged in the surveillance and gathering of intelligence about their adversaries. In 1934, the SD moved from Munich to Berlin, where it established its headquarters in the *Prinz Albrecht Palais*. The move required changes in the organization and work of the agency.[5] One RSHA official whose work had particularly strong influence on these changes was Franz Alfred Six, who served in the agency between May 1935 and March 1943.[6] An expert in journalism at the University of Berlin, Six was initially appointed to head the SD's press division, and from 1939 was charged with the responsibility of researching the opponents of the regime within a department known as Ideology, Research and Evaluation. In this capacity, Six was responsible for developing the library, a museum, and scholarly research.

The Third Reich viewed itself as being threatened by Jews, Freemasons, monarchists, immigrants and political churches, all of which were considered to be enemies and ideological opponents of the regime. Six recommended thorough research of the writings of these opponents in order to be able to better recognize and understand what he referred to as "the spiritual weaponry of ideological opponents." Research of the opponents was portrayed as an historical-cultural hybrid discipline aimed at examining, evaluating, researching and interpreting the regime's ideological enemies, and quickly emerged as a new multidisciplinary field with no boundaries. Its researchers, who consisted of sociologists, linguists, ethnologists, art historians and researchers of the press, formed an elite group of analysts who transformed the field into a dominant scholarly subject. They were provided with financial resources and respect and enjoyed prestigious careers. Six directed this research for a significant period in his capacity as director of Department VII of the RSHA. What set this department apart from similar research institutions, however, was the fact that both organizationally and conceptually, its activities were directly linked to the policy of extermination of the Jewish people.[7]

When the RSHA was established in September 1939, Six proposed a new organizational structure to Heydrich that included a research department that would suit his needs and that would include the research of regime opponents, documentation and library services, and more. His recommendations were eventually fully integrated into the organizational structure of the RSHA, and a number of changes to this end were made between 1941 and 1943. From that point on, Department VII: Ideologies - Research and Evaluation included the following units:

Unit	Unit Heads		Assignments
	1941-1943	1943-1945	
VII Dept. Director	Prof. Franz Six	Dr. Paul Dittel Walter Bruederle	Ideological Research and Evaluation: Administration, manpower, budgets, registration
VII Special Unit		Dr. Günther Stein	Scholarly organization and publications
A	Paul Mylius	Karl Burmester	Collection of Material
A1	Dr. Waldemar Beyer Dr. Günther Stein (until March 1942)	Karl Burmester (until April 1942)	The library
A2	Helmut Mehringer	Karl August Focke	Reporting, translation, and the selection and use of press material[i]
A3	Karl Burmester	Walter Braune	Confidential investigations[ii]
B	Unstaffed	Erich Ehlers	Evaluation[iii]
B1	Unstaffed	Erich Ehlers	Freemasons
B2	Dr. Friedrich Murawski	Dr. Heinz Ballensiefen	Jewry (formerly Political Churches)[iv]
B3	Dr. Horst Mahnke	Rudolf Lewin	Political churches (formerly Marxism)[iv]
B4	Rolf Mühler	Georg Heuchert	Marxism (formerly Other Opposition Groups)[iv]
B5	Dr. Hans Schick	Dr. Hans Schick	Liberalism (formerly research on domestic problems)[iv]
B6	Unstaffed	Rolf Mühler / Horst Kunze	Other opposition groups: pacifism, emigration, etc. (formerly research on problems abroad)[iv]
C	Unstaffed	Rudolf Richter	Archive, museum, and special assignments
C1	Dr. Paul Dittel	Rudolf Richter	Archives[v]
C2	Unstaffed	Rudolf Richter	Museum

Adapted from the following sources: R. Rürop (ed), Topography of Terror, (translated by W.T. Angress) (Berlin, 1998), 78-82 and J. Rudolph, "Sämtliche Sendungen sind zu richten an das RSHA Amt VII 'Weltanschauliche Forschung und Auswertung' als Sammelstelle erbeuteter Archive und Bibliotheken," in M. Wildt, (ed), Nach richtendienst politische Elite und Mordeinheit (Hamburg, 2003), 239-40.

Notes:

i. Unit A2 maintained a collection of German and foreign newspaper clippings, sets of reports issued by all other units, and memoranda written by RSHA agents concerning the activities of "enemies of the Reich." All these were arranged in chronological order by subject.

ii. Unit A3 compiled biographical information from international encyclopedias, telephone books and other published directories.

iii. In Division B, Six consolidated all the units concerned with ideological research regarding the enemies of the Reich. The function of these units was to evaluate the confiscated spoils and propose courses of action to other units. Their goal was to uncover networks of influence and resistance consisting of Jews, immigrants, Freemasons, churches and other sects, with the ultimate aim of establishing control over their leaders and taking them into custody as enemies of the Reich.

iv. "Formerly" here refers to the period between 1941 and 1943.

v. This unit maintained the political archives of the SD and hostile states.

In January 1936, Himmler ordered the establishment of a central scientific library within the SD, marking the first usage of the term "central library" in this Nazi institution.[8] According to the plan, the library was intended to hold all political literature that was "damaging and undesirable," including that written by enemies of the regime. It was to include Judaica and Hebraica; liberal, pacifist, and Marxist literature; the writings of psychoanalysts; literature in the fields of sexology and anthroposophy; Freemason literature; and criticism of Nazism. However, the project became operative in practice only in 1939, with Six's consolidation of the Jewish libraries within one unit, as we will see below.

When the Nazis came to power in 1933, the police, the legal system, and Nazi Party institutions in Bavaria began confiscating all publications and archives of the institutions, organizations and parties that had been declared illegal. Among the first materials confiscated were the archives and libraries of the Freemasons, which were transferred to the SD in Munich and later moved to Berlin, where they in fact constituted the foundation of the SD's library. Throughout the Reich, Freemason lodges were ordered to transfer their libraries and archives to Berlin. Their large building at 12-13 Emser Street in Berlin was also appropriated and used as a warehouse to store the crates of books sent by the lodges. The purpose of these measures was to create a systematic and comprehensive collection of Freemason literature. In 1936, the size of this collection was estimated at approximately 600,000 volumes, from which approximately 50,000 copies were taken to create reference collections for the various units of the SD.[9]

During this period, these collections could still not be considered a central library. The following years witnessed the development of additional collections, such as a legal library, divisional reference collections and a library of prohibited literature. The existence of these collections is reflected in the structural and organizational plans of the SD and the RSHA. Some libraries operated as separate units or were maintained jointly as parts of museums, but all such libraries were related to specific research functions.

The library of looted Jewish books – the *Judenbibliothek* – was established in 1937, along with the "securing" of the libraries of 150 branches of B'nai B'rith. A large part of these collections were destroyed, and only a selection of 1,800 books, selected by librarian Günther Stein, were sent off to Berlin. The following year, the collection absorbed an additional 3,600 books from the Leipzig *Delitzschianum Judaicum*

Institutum. This was the core of the Jewish library, which now began to evolve into a large, significant collection.

The Reich's annexation of Austria in March 1938 provided a test case and a site for experimentation in which, for the first time, the SD could assess the work of the *Einsatzkommando*. Five professional librarians from the SD library took part in missions related to the closure of institutions that had libraries and the reviewing and "securing" of their collections. Stein was charged with supervising the shipments from Austria to Berlin. Approximately 130 tons of books and archival documents were sent off to the German capital from Vienna in thirteen shipments. The important collections of Austrian Jewish organizations such as the *Israelitische Allianz, Hebräeisches Pädagogium* and *Union Österreichischer Juden* were consolidated in approximately 200 crates. To cover the logistical expenses of moving the libraries to Berlin, Adolf Eichmann provided the SD with 100,000 marks from the budget of the Jewish Emigration Fund. An additional 150,000 marks were transferred from the same fund in Vienna for the establishment of the Jewish library, and another 12,000 marks were provided from confiscated Jewish funds.

By the end of the first year of annexation, the library's collection had expanded to 85,000 volumes.[10] During *Kristallnacht*, which took place later that year on November 9, 1938, Jewish synagogues were destroyed, Jewish institutions were closed and the homes of Jewish families were damaged. However, the rioters were warned ahead of time to refrain from damaging the offices, libraries, and archives of the organizations, which the police confiscated and protected from the demonstrators. Himmler ordered that the libraries' catalogs be reviewed, and Stein prepared a list of 71 Jewish libraries that were designated for confiscation which was subsequently submitted to Heydrich. According to their catalogs, these libraries held a total of approximately 300,000 volumes. The list did not include the collections of books that had already been confiscated, such as the libraries of B'nai B'rith or the library of the Central Union of German Citizens of Jewish Faith (*Centralverein deutscher Staatsbürger jüdischen Glaubens*). The list specified sixteen libraries that held 200,000 books of special value, including the libraries of the Jewish communities of Berlin, Frankfurt am Main, Munich, Vienna, Dresden, Hamburg and Hannover; the libraries of synagogues in Breslau, Hamburg and Cologne; and the libraries of the rabbinical seminaries in Berlin, Breslau and Frankfurt.[11]

There is no doubt that the events of *Kristallnacht* served as a significant catalyst for the development of the central library of looted books. Planning and implementation was controlled by Six and his librarians. Both Six and Dittel, his successor, argued to Heydrich that centralized control of the Jewish libraries was of decisive importance from the perspective of the SD. The libraries, they claimed, were an important means of researching Judaism and Jewry and of achieving the goals of the Nazi regime vis-à-vis the Jews. To this end, they advocated confiscating and "securing" the books, as opposed to destroying them, and using them to establish a central Jewish library (*Jüdische Zentralbibliothek*) under the auspices of the SD. They estimated that transferring and shelving the books would cost approximately 70,000 marks, an outlay that was justified, they claimed, by the books' real ideological and political value, which outweighed the expense it would involve. In this manner, they stressed the importance of consolidating the Jewish libraries as a mission that would serve both the party and the state. Heydrich agreed to the recommendations of Six and Dittel and officially authorized their implementation in March 1939.[12]

The first libraries confiscated and transferred to the SD immediately following Heydrich's authorization included the library of the Central Union of German Citizens of Jewish Faith (15,000 volumes); the books that had been collected by the German Friends of the Hebrew University for the Jewish National and University Library in Jerusalem (24,000 volumes); the private library of Jewish author Leon Feuchtwanger, which included one hundred rare books; and the inventory of Seemann's bookstore, which was estimated at approximately 100,000 volumes, some of which were spared shredding.[13]

At the onset of the Nazi invasion of Poland, the collections of the Polish Jewish libraries were shipped to Berlin in six train cars. Most of the synagogue collections, the collection of the large Jewish library on Tlomackie Street in Warsaw, the Borochov library and various additional collections were deposited in the warehouse in the Freemason lodge at 13 Eisenacher Street. At Six's request, the building had been seized and specially renovated to fulfill its purpose as the office and storage facility for the looted collections. Following Germany's conquest of the Baltic States, Jewish libraries there were also confiscated and sent to Berlin. They included the archive and library of Simon Dubnow and the collections of the Jews of Tallinn (Reval). Six consolidated the most important collections within Unit A1 of his department, which was also responsible for documentation. These two buildings, one on Emser Street and the other on Eisenacher Street, stored the looted collections and housed the offices of Department VII A. This marked the beginning of the implementation of Himmler's 1936 directive, and the Central Library became a reality.

Six also initiated large-scale book exchange agreements, such as the exchange concluded between the RSHA and the Prussian State library in Berlin. For example, in 1937, Six asked this library to grant the RSHA volumes of Russian literature, offering books from the looted collections in the Freemason lodges in return. By means of such exchange agreements, the scholarly and public libraries in Germany and Austria received many books that were formerly part of collections looted by RSHA personnel.

A number of assessments have been advanced on the question of how many books were actually held by the library.[14] Estimates made at the time are not detailed and relate to the overall number of books in the RSHA libraries alone, and not in the other collections. Stein, the library's director at the time, estimated that in 1941, the library held between 500,000 and one million books. Ernst Grumach, who headed the group of forced laborers assigned to the library, claimed that it was impossible to know the exact number of books, but that they most likely totaled between two and three million. In contrast, Paul Dittel, who was appointed the library's director in 1943, estimated that, at that time, the library held only 200,000-300,000 volumes. Grumach's numbers appear to be more realistic, as they also included the "secured" and confiscated collections of the Freemason lodges and the Christian sects, Marxist literature, and the Jewish collections.

II. The Assignment of Jewish Forced Laborers and the Evacuation of the Library

To solve the problem of the lack of professional personnel required to handle the enormous quantity of books collected, Six proposed recruiting between six and eight Jewish librarians as forced laborers, whose salaries would be paid by the Central Office for Jewish Emigration (*Reichszentrale für jüdische Auswanderung*). The group was to constitute a unit under SS command (*Bibliothekskommando*) that would operate out of the Jewish community building in Berlin. Heydrich, however, insisted that the work be carried out nine hours per day in the stacks on Eisenacher Street. The forced laborers worked in isolation from the other library employees and were subject to constant supervision and thorough inspections. They were also required to maintain complete confidentiality regarding anything pertaining to their work.[15]

Waldemar Beyer, the library's director at the time, was charged with selecting the candidates and reviewing their résumés. The names of women and candidates with families who had declared their desire to emigrate were removed from the list of candidates, as were all former employees of state libraries. These recommendations were submitted in March 1939 but were implemented only in September 1941, when it became apparent that the shortage of storage space, shelves, and skilled personnel had brought the classification and cataloging of books to a standstill.[16] Now, the Federal Union of Jews in Germany was charged with the task of supplying eight Jewish librarians. In October 1941, the candidates were summoned to Eichmann's office, where they were inspected by RSHA librarians Günther Stein and Walter Kellner.[17]

The laborers began their work in the Eisenacher Street building on October 3 under Kellner's supervision.[18] Ernst Grumach was appointed *Oberjude* (head of the work group) and given almost unlimited managerial authority over the workers. He was also responsible for solving their personal problems, which included situations that were sometimes delicate and dangerous.[19] For example, by increasing the number of forced laborers at work in the library, Grumach made great efforts to remove his workers and their families from the deportation lists, although these efforts were largely unsuccessful. He consulted with his colleagues regarding fateful decisions and imbued them with a sense that fate had assigned them a unique mission: preservation of the confiscated Jewish libraries and their reorganization into one large central Jewish library.

The forced laborers engaged not only in professional tasks such as classification, cataloging and shelving, but also in difficult physical tasks such as unloading crates, packing and loading crates onto trucks, cleaning the library and the commanders' official quarters and making repairs in their apartments, repairing the roofs of the building after bombings, building bunkers, and, in the winter, shoveling snow. The workers were subject to pressure, insults, harassment and mockery. There were also instances of informing, interrogations, trespassing and persecution on the part of SS men, which were met with severe punishments such as immediate deportation to an extermination camp.[20] On March 9, 1943, following the *Fabrikaktion*, the Jewish work group was disbanded and its members were sent to Auschwitz.[21] Of the group's twenty-four members, the only survivors were Ernst Grumach and Bethold Breslauer, who were married to non-Jewish women.

After this wave of deportations, the Nazis enlisted a second group of twenty-five Jewish forced laborers. In Grumach's opinion, this group was not as effective as the first group because most were not proficient in the Hebrew language. This group, however, survived the Holocaust. Their salaries, the costs of shipping the books and other expenses were paid for using an illegal bank account placed at Six's disposal, to which Eichmann transferred substantial sums from the funds of the offices then dealing with Jewish emigration. In other words, the expenses were paid for by the victims themselves.

Following the heavy bombings of Berlin in August 1943, an order was issued to move the historical material held by Department VII to safe locations. With the danger of death hovering over them, the Jewish laborers worked fourteen to sixteen hours a day removing books, loading trucks and guarding them at night. By the summer of 1944, a large portion of the books in the RSHA library had been moved to safe locations.[22] The scope of work in the library was therefore reduced, and most Jewish forced laborers were left without work and may have been placed in other units of the RSHA. In any event, they were not sent to the extermination camps, apparently out of concern that news of their deportation would encourage escape attempts by other forced laborers employed by the RSHA.

Most of the Hebraica books were sent to the Theresienstadt concentration camp, where prisoners fluent in Hebrew were entrusted with classifying and cataloging them. Fifty-eight crates of books holding approximately 60,000 volumes were delivered to the camp library at Theresienstadt, which was directed by Rabbi Benjamin Murmelstein. By March 1944, the library had recorded and cataloged some 28,000 Hebraica books. Between 1945 and 1946, these books were moved, along with the ghetto library, to the Jewish Museum in Prague, which also received books from Castle Niemes in Bohemia. A number of Jewish collections that the Germans had not evacuated to other safe locations remained in Berlin. Grumach later claimed that in May 1945, thousands of books from the Jewish collections were still located in the building on Eisenacher Street.

When the Red Army conquered Berlin, the Soviet military appropriated Department VII, Unit Al but left the gates of the warehouses without a secure lock. As a result, they were the site of massive looting, and many books ended up in the hands of local city residents, and especially book dealers. A few weeks after the fall of the city, the Communists appointed a trustee for the library and removed books that they regarded as useful for their own purposes. When the US Army reached Berlin in July 1945, Grumach convinced an American officer of the importance of caring for the remaining books. The US Army confiscated them and transferred them to the Offenbach Archival Depot, which immediately began a process of restitution. In March 1947, forty-seven crates of books–including books from the Freemasons collection, Jewish manuscripts and private collections–were shipped to Moscow.

III. Organizational, Managerial, and Professional Aspects

Upon his appointment as director of the library in 1942, Karl Burmester outlined his plan for the library's reorganization.[23] His proposals were reflective of the poor condition the library was in. The major problem was the lack of professional personnel, wooden crates for packing books and metal shelves for unpacking them. That year, most of the books remained in crates scattered among the basements and halls, and only a small percentage were actually examined and shelved. The rest were left in storage until the arrival of metal shelves, which were confiscated from monasteries in France and Germany and sent to Berlin.

Burmester's proposal also shed light on the working procedures in place in the library. The books were supposed to be arranged by subject and two catalogs were planned: an alphabetical union catalog and a subject catalog. However, as the process of the classification of the books was a long-term project and as professional personnel were in short supply, it was decided to give preference to establishing the union catalog. In the end, only a small number of books were cataloged using an abbreviated cataloging process. Only books published during that period (i.e. during the War) were cataloged properly, in accordance with the cataloging system known as the "Prussian Instructions" that was customary in the country at the time.

At one point, it was proposed that all books without use be destroyed. Eventually, a large number of books were taken away from the library, although no information has thus far been found indicating the number of books that were destroyed. Burmester also resolved to transfer entire collections to other institutions. For example, the collection of the Hamburg Jewish community was moved to the city's municipal museum.

The library had no reading room, lacked regulations for use and borrowing, and was used primarily by the personnel of Department VII. The library's general unpopularity was due to its inaccessibility and to its many defects which made extensive research work nearly impossible.[24] According to Burmester's plan, the books would be shelved by subject matter, and access would be limited to library personnel only. The users of the library would locate the material they needed in the catalogs with the assistance of the librarians, which was consistent with the closed-shelf policy that was customary in German libraries at the time.[25] This plan, however, was never actually implemented.

Organizationally, the RSHA library of Department VII, Unit A1 was divided into three large subject-based categories, into which the confiscated books were integrated (along with new books purchased from the RSHA budget): 1) the German Reich and reference literature; 2) regime opponents; and 3) foreign countries. The books in the "opponents" category were further broken down into four divisions: a) church and pseudo-religious groups, such as theosophy; b) the Freemasons; c) Marxism; and d) Judaism. Of all these collections, only a portion of the books on Judaism and Jewry were actually arranged and shelved. The books were arranged according to alphabetical order under the following subjects:[26]

1. General and Jewish reference literature, divided into the three following areas: general and Jewish reference books, Jewish periodicals, and Jewish works in series and commemorative volumes.
2. Hebraica (material written in the Hebrew language).
3. Judaica, subdivided into the library of the Jewish community of Berlin, the library of the Jewish Theological Seminary of Breslau, the library of the Central Association of German Citizens of Jewish Faith, and the private library of Professor Sigmund Seligmann of Amsterdam.
4. Books on non-Jewish subjects written by Jews.
5. Jewish and Hebrew manuscripts from all confiscated libraries.
6. A collection of 60,000 pamphlets, reports, and printouts on Jewish-related subjects.

A large number of books (multiple copies) were removed from the Jewish collections and given to Nazi institutions such as Heydrich's office library and the SD Wannsee Institute. Others were given to civilian institutions, such as the Prussian State Library in Berlin and the Bavarian State Library in Munich. Some of the books that were categorized as political and ideological literature of regime opponents were transferred to the Institute for Research into the Jewish Question in Frankfurt am Main. In effect, the confiscated libraries of the Jewish institutions underwent a process of significant dispersal and fragmentation. The winnowing process completely destroyed the organic unity of entire libraries that had been systematically built-up over the course of generations, and not a single library was preserved as a complete whole.

Various agents competed with one another over the plundered collections, and their relations with each other were characterized by a degree of unstable cooperation. First selection from the looted books was typically reserved for the RSHA. Six demanded exclusive RSHA use, although he and Dittel, his deputy and successor, both recognized that the library of the Institute for Research into the Jewish Question in Frankfurt am Main also maintained important material that was likely to serve research interests of the RSHA.[27]

IV. The Individuals Involved in the Looting

Those who brought about the establishment and development of the library of looted books–those who were in charge of the institution and set its policies and its modes of operation–also played an active role in the looting process. Their reports provided accounts of their looting trips abroad to the East and the West. The central figure and the most influential personality in terms of making policy determinations within Department VII, was Franz Alfred Six. The son of an upholsterer, Six was born into a lower- middle-class family and was not expected to pursue higher education. Nevertheless, he studied journalism and political science at the University of Heidelberg and in 1934 received a doctorate for a dissertation dealing with Nazi political propaganda during the party's struggle for political power. He became a member of the *Hitlerjugend* (Hitler Youth) in 1929 and of the Nazi Party and the SA in 1930.

In 1935, he began working in the SD's journalism department and was later appointed director of Department VII, which engaged in the research and evaluation of the ideologies of regime opponents. Six left the RSHA in 1943 and was appointed head of the Political-Cultural Department of the Foreign Ministry, with the rank of ambassador. Along with his work in the RSHA, Six also developed an academic career in the field of journalism at the universities of Königsberg and Berlin between 1935 and 1943 and was promoted to the rank of full professor by Hitler himself. At the University of Berlin, Six served as dean of the Institute of Foreign Studies. At the war's end, Six tried unsuccessfully to evade the Allied military authorities by changing his name. He was arrested and sentenced to twenty years in prison, but was ultimately released in 1952 and, until his death in 1975, worked as an economic and marketing consultant for various companies, primarily Porsche.[28]

Six's involvement in the Nazi activities that preceded the genocide of the Jews has yet to be sufficiently assessed. Indeed, today we know that Six had considerable influence on the execution of Nazi policy regarding the Jews and the logistics of Jewish persecution. He supported Adolf Eichmann, who in 1938 was sent by him to Vienna, where he developed a rational bureaucratic model for the expulsion of Jews through the Reich's Central Office for Jewish Emigration – a model that was implemented later in other countries.

Centralized control over the Jews and their expulsion were perceived to be the ultimate activity of Six's department. As the SD's press expert,

Six's main concern was to develop a record of the intellectual writings of the Nazis' ideological opponents, based on the confiscated books from the libraries of the Freemasons, the Catholic Church and other collections. Opponents were linked to one another using diagrams and maps, and in this manner groups and individuals were linked to a conspiracy masterminded by the most formidable enemy of all – the Jews. Using information on prominent Jews in all realms of life both inside and outside Germany, they were investigated and arrested and their organizations were destroyed.

The pogroms and the wave of terror that swept Austria following its annexation were prepared and planned in the "Jewry" department that was controlled by Six, who inspired a militant atmosphere in the units he supervised. Six was also involved in various *Kommando* activities, including heading a *Kommando* unit in Vienna in March 1938 and organizing the confiscation of Jewish libraries and archives. Paul Dittel, Six's deputy and eventually his replacement, described Six's nature and personality:

> Six is a fanatical Nazi and he is not the type that can be converted. He may become aware of his own mistakes but he would never admit them… He is one of those who compensate [for] their inferiority complex by a self-assured and boastful demeanor. He tyrannized over his staff wherever possible, although he sometimes protected those who enjoyed his special confidence and was not unforgiving. He is always full of fresh, if impractical ideas. At the beginning he had little respect for scholarly research although he was still far too "intellectual" for Kaltenbrunner`s liking… In conversation he was outspoken and even brutal, but frequently not to the point. A parvenu of unrestrained ambition, to whom his own career meant everything.[29]

Paul Dittel ranked second to Six in the bureaucratic hierarchy of Department VII, serving first as Six's deputy and, beginning in 1943, as its acting director.[30] In 1938 and 1939, he was part of a *Kommando* team that took part in the looting of libraries and archives in Austria and Poland. In 1942, along with his duties as Six's deputy, he was also made responsible for the museum, the Freemasons, scholarly research, and the library. The synagogue building at 37 Münchener Strasse, which was confiscated for RSHA use in 1941, became Dittel's residence in August 1943.

During his interrogation by British intelligence after the War, Dittel claimed that from a National Socialist perspective his department, which was also responsible for the library, had been a failure from the beginning. Its purposes were never defined, and instead of engaging in political intelligence, its staff engaged in academic historical research using historical methodology that the Nazis had rejected. According to Dittel, the research carried out within the department was never appreciated because it contributed nothing to the war effort and was therefore found to be unnecessary. When considering these claims, however, it is important to remember that during his interrogation by British intelligence, Dittel had a vested interest in playing down the importance of his position, his work, and his department as a whole. The introduction to the report on his interrogation indicates that Dittel had been very impressed by early Nazi activity but that his views changed once he saw that his work was not appreciated and because of his poor relationship with Six, his immediate superior. His British interrogators emphasized that he completely rejected the Nazi worldview and gave them his complete cooperation. They described him as a non-aggressive, unpretentious man who was well qualified for research and who repeatedly expressed his shock at the corruption and cruelty of the Nazi regime.

Ernst Grumach's testimony reveals a different side of Dittel's character. Although the forced laborers were not in constant contact with Dittel, Grumach explained, he nonetheless succeeded in making everyday life difficult for them. Dittel was a brutal alcoholic who regularly used his fists to threaten the Jewish forced laborers.[31] He also warned Grumach to keep close watch over his workers, as, in his view, "one Jew more or less is of no consequence."

In 1936, Six initiated a process of professionalization within the SD library that involved new appointments throughout the library.[32] The library's specialist on the Freemasons was Dr. Ortwin Dietz, a lawyer who had worked in the University of Leipzig library and moved to the library of the University of Greifswald in 1939. An early Nazi, Dietz also worked in the SD library from 1936 to 1939. Werner Kaempf, also an early Nazi, had worked in the SD library since 1935 and was the library's specialist on Marxist literature. To direct the library, the Nazis appointed Dr. Waldemar Beyer, who had studied at the University of Leipzig, worked in its library from 1936, and moved to the SD library after passing the state librarian certification examinations. At the outbreak of the War in September 1939, he joined the air force,in which he served with the rank of captain but also remained responsible for the library until 1942, when he moved to the University of Strassburg. During his interrogation by the British, Dittel, Beyer's immediate superior, recalled that Beyer was never present at the library during the War. In fact, the library had lost all its senior librarians by the beginning of the War, and only two of the library's ten subject specialists remained.

As the library's specialist on Jewry and churches, the Nazis appointed Dr. Günther Stein, who had studied at Leipzig's public library school and worked at the Reading and Literature Institute established by Walter Hofman. Stein also studied medieval history at the University of Leipzig where he was a student of Professor Helman, who was Jewish. Until 1936, Stein worked on organizing the local Nazi Party library in Leipzig, and in September of that year he began working at the SD library. In 1939, he was appointed deputy director, a position he retained until 1942. In a report on the library, Stein wrote that the substantial expansion of its collections and the shortage of professional librarians significantly reduced the library's professional organization and its ability to handle books. According to Stein's assessment, the library held between 500,000 and one million volumes. In his report, Stein acknowledged that the library was actually a failure. Shortly after his report he was dismissed.

Stein also took part in the looting of the Jewish libraries. Among other things, he was responsible for the confiscation of the property of Dr. Gottfried Kurnwald, including his library, which he sent off to Berlin. Kurnwald had on a number of occasions served as an advisor to the Austrian government .When the Nazis burst into his home he committed suicide. Stein was also involved in the confiscation of the inventory of the publishing houses of Querido and Allert de Lange in Holland, which were known for their publication of exile literature (*Exilliteratur*). According to Dittel, Stein's superior, Six viewed Stein as a pedant and constantly interfered with his work. Six's interference and the humiliation he caused Stein prevented the latter from carrying out professional and systematic work in the library. Stein was extremely offended by Six's insulting treatment, which may have been one cause of his long illness and extended vacations. When he recuperated, he was transferred to the department's political churches unit. As a result of his experience, Stein grew disenchanted with Nazi ideology and began to be critical of it. Grumach described Stein's personality as shrouded in lack of clarity and at odds with his surroundings. He was fair to the Jewish forced laborers and even tried to help them, to the degree that this was possible.

After Stein's departure in 1942, Karl Burmester was appointed as the library's director.[33] Despite his lack of academic education and training as a librarian, Six appointed him as his personal assistant and, subsequently, as the director of the library. Burmester served as the library's last director between 1942 and May 1945. Six believed that although he would be less academic than his predecessors, he would handle library materials more efficiently. During his interrogation, Dittel claimed that Burmester was one of the library employees who was critical of Nazi ideology. Grumach, however, described Burmester as a man who had an unrestrained personality and was capable of doing anything. In Grumach's own words, Burmester was "e*in Schweinehund ohne gleichen*"-- a real pig.

Dr. Walter Kellner studied theology and education at the universities of Erlangen and Münster and served as a priest during the first half of the 1930s. In 1936, he left the Church and started working for the SD, where he served as a subject specialist and was responsible for the Jewish collections in Unit A1. At the end of the War, he was tried by a Russian military tribunal and sentenced to twenty-five years forced labor in the Soviet Union.[34] Six had Kellner take part in the planning of an academic conference of experts on solutions to the Jewish question. Kellner's lectures pertained to Jewish settlement in Palestine and the Jewish influence on education in Germany. He was also involved in the looting of libraries in the East and West, including the confiscation of the private Judaica collection of Professor Sigmund Seligmann in Holland and its transport to Berlin. This library survived the Holocaust and, after the War, was returned to his son Izhak Seligmann, a professor of biblical studies at The Hebrew University of Jerusalem.

Kellner was the direct supervisor of the Jewish forced laborers in the library, and in this capacity he appointed Ernst Grumach to head the group. Dittel regarded Kellner as a good and extremely knowledgeable worker. Even though he was a member of the Nazi party, he voiced criticism of the party and was not Antisemitic. Dittel characterized him as unstable. Grumach, who had contact with him on a daily basis, also described him as having an extremely strange personality, as a "degenerate offspring" of one of Germany's oldest Catholic families, and devoid of any operational abilities. His one desire was to enable Grumach to do the tasks assigned to him to the best of his abilities so that he could later take credit for Grumach's achievements as if they were his own. To his credit, he tried to help his forced laborers from being deported to the extermination camps, and was ultimately successful in saving one of his workers from immediate transport.

But Kellner also had his vices, including an addiction to morphine. To finance his morphine habit, he would sell books from the looted collections to booksellers. He was not the only addict in the RSHA. Other addicts within the agency were exposed when they came into contact with drug dealers who had been planted in Denmark by British intelligence, and who provided the British with significant information regarding RSHA activities. This was their source of information regarding Kellner's as well as Grumach's and the RSHA`s activities.

In addition to the abovementioned personnel, others worked in the library at various points. According to Burmester, in 1942 the RSHA library employed thirteen librarians, four interns from the Prussian State Library in Berlin, and two assistants. Dittel's report refers to a number of additional employees, such as Karl Engelmann, a veteran Nazi party member who worked under Burmester sorting books, and Walter Brüderle, who served as head of the administrative office and who, according to Dittel, was an efficient but embittered worker. Others were Frau Durow, a small, inconspicuous woman who was a member of the Nazi party. She was in charge of cataloging Marxist literature. Heinz

Fichtner, also a member of the Nazi party, but "not one hundred percent Nazi," performed technical tasks. Möckel, a Party member who did not hold Nazi views and joined the Party late, engaged in book purchasing, and was regarded as an efficient and hard working man. Gerhard Schmidt was employed in sorting and classifying books. He was a long-time Party member, and, according to Dittel, had a moderate outlook and an embittered personality. In addition, Frl. Tübekke, who was apparently not a Party member and was indifferent to politics, cataloged the Jewish books. Grumach's testimony refers to another library employee: Franz Zimmerman, an academic librarian from the University of Halle.

V. Conclusion

Several of the documents indicate that the purpose of the Central Library and its functions were defined and perceived differently by the various individuals who were in charge of it.[35] Whereas Burmester sought to develop the library into a large reference library in which books would be shelved by subject, Dittel regarded it as an archival collection for research purposes, in which each book and pamphlet constituted an archival document. According to Dittel, the library was not intended to serve as a reference collection for RSHA personnel, as collections of this kind already existed in the various units of the agency.

Dittel's view was undoubtedly closer to reality than Burmester's. The vast number of Jewish-owned books that had been consolidated within Unit A1 of Department VII constituted a central library only in theory. Although in scope it represented one of the largest collections in the Reich, it was actually a conglomerate of confiscated library collections that served as a basis for Nazi research on the ideology of Judaism and Jews. The products of the research undertaken – such as the lists regarding individuals wanted by the Gestapo, reports on the confiscations, guidelines for additional looting operations, evaluations and biographies – all served a purpose with regard to carrying out anti-Jewish activities and the ideological and ritual training of SS members. In this way, they also contributed to the de-legitimization of the Jews, which was a necessary precondition for their extermination.

Historian Jörg Rudolph was correct in his assertion that the librarianship practiced in the RSHA library was indistinguishable from that applied in the civilian libraries throughout the Reich.[36] The archival documents suggest that there was a constant desire to perform professional work and to adapt the library to fulfill its role as a research library. However, time constraints, the shortage of storage space and professional personnel, the exposure to heavy bombings and the complicated process involved in evacuating the books to safe locations were all impediments that dictated many of the failures in the library's daily work. As a result, entire collections that had been accumulated systematically for generations were completely dismantled. From their components, new collections were assembled that met their needs. This reorganization involved the work of forced Jewish laborers, who regarded their task as a mission, and believed that they were working to create a large central Jewish library. But their sense of mission was false, as in the course of the reorganization many books were transferred to other institutions throughout the Reich, and thousands of books, including prayer books and rabbinical literature, were shredded.

Although Department VII did not contribute to the Reich's war effort, and although the biographical information and activities of its personnel may not be particularly interesting, tracing the life stories and the work of these individuals supports the conclusions of Michael Wildt's study of the leadership of the RSHA.[37] As the archetype of criminals, RSHA leaders differed significantly from other criminals within the Nazi regime. They were unique in their homogeneity, their academic training and their relative youth. Furthermore, academics such as Stein, Six, and Beyer did not merely plan their crimes at their desks in their offices; they also carried them out in practice, as part of *Einsatzgruppen* or *Kommando* units. There was an underlying connection between Nazi ideology and the RSHA as an institutional center of racial and anti-Semitic persecutions and for the planning of genocide. These characteristics were prominent with respect to the activities of most of those who looted the Jewish libraries and who also worked at the library. Indeed, their activities contributed to the process of the legitimization of genocide.

Notes

* This article is a revised and updated version of "The Library of the Reich Security Main Office and Its Looted Jewish Book Collections," *Libraries and the Cultural Record* 42,1 (2007), 21-47.

1 On the looting and restitution of Jewish cultural property, see: Dov Schidorsky, *Burning Scrolls and Flying Letters,* (Jerusalem, 2008), 150-291, 343-467 (Hebrew).

2 Ernst Grumach (1967-1902) was born in East Prussia and studied classical philology at the University of Königsberg, where he began to teach in 1930. Later, he taught Latin and Greek at the Higher Institute of Jewish Studies in Berlin until its closure by the Nazis in 1941. It was then that he was assigned to forced labor in the RSHA library. He eventually survived the Holocaust due to the fact that his wife was not Jewish. Most of his writings focused on the field of classical culture, and he served as an editor of a volume of the complete writings of Aristotle. For Grumach's testimony regarding his work as a forced laborer see: "Gedächtnisniederschrift über die Besprechung am 3. März 1960, mit Herrn Professor Dr. Ernst Grumach in seinem Hause in Berlin-Wilmersdorf, Berliner Str. 60," Central Archives for the History of the Jewish People (CAHJP), Jerusalem P/205. See also: Dov Schidorsky, "Confiscation of Libraries and Assignments to Forced Labor: Two Documents of the Holocaust, " *Libraries & Culture* 33 (Fall 1998), 347-388.

3 Bericht Karl Burmester vom 22 April 1942," in Andrezej Męzynski, *Kommando Paulsen: Organisierter Kunstraub in Polen, 1942-1945* (Köln, 2000), 85-91 (herein, "The Burmester Report").

4 Report on Information obtained from PW CS/2294 SS Ostubaf. Dr. Paul Dittel," September 13, 1945, Public Record Office, London WO 208/2620 (herein, "The Dittel Report").

5 See Table below.

6 On Franz Six see Lutz Hachmeister, *Der Gegnerforscher. Die Karriere des SS Führers Franz Alfred Six* (Munich, 1998), 30 ff. See also: The Dittel Report, Appendix 3, Section 78.

7 Jan Björn Potthast, *Das Jüdische Zentralmuseum der SS in Prag: Gegnerforschung und Völkermord im Nationalsozialismus* (Frankfurt am Main, 2002), 21, 28-29.

8 The Gestapo sent a circular containing Himmler's directive of January 22, 1936 to all police stations. See Werner Schroeder, "Strukturen des Bücherraubs: Die Bibliotheken des RSHA: Aufbau und Verbleib," Lecture manuscript, Weimar, Germany, September 9, 2003, 4.

9 Werner Schroeder, "Bestandsaufbau durch Plünderung – Jüdische Bibliotheken im Reichssicherheitshauptamt." Paper presented at the Tagung Wiener Stadt-und Landesbibliothek, "Raub und Restitution in Bibliotheken," Wien, April 23-24m 2003, 1-2.

10 On the fate of the Jewish-owned libraries and collections in Austria, see: Evelyn Adunka's study *Der Raub der Bücher* (Wien, 2002).

11 "Fassung und Sichtung aller sichergestellten jüdischen Bibliotheken," February 1939, Bundesarchiv Berlin R58/6424, 354.

12 "Bericht betr. Zentralisierung der Judenbibliotekhen vom 1 bzw 5.6.39," Bundesarchiv Berlin R 58/6424, 431-435. See also Schroeder, "Strukturen des Bücherraubs," 10.

13 Schroeder, "*Bestandsaufbau durch Plünderung*," 4.

14 Estimates regarding the number of books assembled in the RSHA library are based on the following documents: Günther Stein, "Bericht über VII A 1 an den Vertreter des Amtschef VII, Berlin 9.4.41," Bundesarchiv Berlin R 58/5407; The Dittel Report, 9. See also: Schidosrky," Confiscation of Libraries…", 358.

15 Jorg Rudolph, "Sämtliche Sendungen sind zu richten an ... das RSHA Amt VII 'Weltanschauliche Forschung und Auswertung' als Sammelstelle erbeuteter Archive und Bibliotheken," in Michael Wildt (ed), *Nachrichtendienst politische Elite und Mordeinheit: Der Sicherheitsdienst des Reichsführers SS* (Hamburg, 2003), 228-229.

16 Schroeder, "Strukturen des Bücherraubs," 12.

17 For more information on these librarians, see section IV of this article.

18 Grumach's testimony includes a list of the names of the forced laborers. See Schidorsky, "Confiscation of Libraries… ", 370-374.

19 The sudden disappearance of one of the forced laborers who found refuge abroad embarrassed Grumach and placed him in a delicate and dangerous situation in relation to his superiors. See Grumach's letter to Wiener, November 11, 1945, CAHJP P 205/19. See also the testimony of Ludwig Misch, undated manuscript, CAHJP P/205/17b.
20 Grumach's testimony describes the forced laborers' difficult working conditions. See: Schidorsky, "Confiscation of Libraries…", 361-369.
21 The *Fabrikaktion* of 1943 was the last raid on the Jews of Germany. During this *Aktion*, the Germans raided stores and factories with the aim of arresting Jewish employees who had hitherto been exempt from expulsion to the East. The names of some of these forced laborers are listed in Rudolph, "Sämtliche Sendungen…," 234.
22 Safe locations included palaces in Bohemia, Silesia, Bavaria, and Thuringia. A detailed list of these sites is found in Grumach's testimony in Schidorsky," Confiscation of Libraries…, " 364.
23 The Burmester Report, 86-88.
24 The Dittel Report, Section 39.
25 The library holdings (books and periodicals) were stored in a closed area to which readers did not enjoy direct access. Library users had to submit their requests to the library personnel at a counter. Then the requested material was retrieved from the stacks.
26 Details regarding the subject areas are provided by Grumach. See Schidorsky, "Confiscation of Libraries…,", 363-364.
27 Rudolph, "Sämtliche Sendungen…," 225.
28 See Hachmeister, 167, 177, 192-194, 218, 229, 235-238.
29 The Dittel Report, Appendix 3, section 78
30 For more on Paul Dittel, see Michael Wildt, *Generation des Unbedingten* (Hamburg, 2002), 368-369, 933; Hachmeister, 170,199, 225; The Dittel Report, 2 and Appendix 1.
31 Schidorsky, Confiscation of Libraries…," 368-369.
32 See Rudolph, "Sämtliche Sendungen," 316-317; Hachmeister, 197; The Dittel Report, Sections 24-33.
33 On Karl Burmester, see Wildt, *Generation*, 369, footnote 250; The Dittel report, Sections 24,33.
34 On Walter Kellner, see Rudolph, "Sämtliche Sendungen," 230; The Dittel Report, Section 36.
35 The Burmester Report, Bericht, 86; The Dittel Report, Sections 24-33.
36 Rudolph, "Sämtliche Sendungen," 216.
37 Wildt, *Generation*, 860-864.

Matityahu Mintz

The Soviet State and the Plight of the Jews during the 'Classified Period' – September 1939- June 1941 – Rescue or Abandon?

"When the Red Army began its campaign of liberation from the Volga to the Bug in the winter of 1942-1943, there were hardly any Jews left to be rescued: virtually all the Jews who had stayed on in the German-occupied zone were massacred. Only a relatively small number, totaling not more than a couple of thousand, was rescued, and that was due to the magnificent courage of local residents in those areas. Several tens of thousands escaped to the forests to join up with the partisans (in the Baranovich region in Belarus, Jews accounted for some 12 per cent of the partisans, and in the adjacent Lida region – 25 per cent). Then there was a substantial number of Jews who fought in the ranks of the Polish partisan army [the Armia Krajova, or 'Motherland Army'], while others were even to be found in the ranks of the UPA – Bandera's Ukrainian rebel forces (Ukrainska Povstanchia Armia). What was Stalin doing here?

The most delicate way to describe the attitude of the Soviet government, and particularly that of the People's Leader, to the massacre of millions of Jews would be icy indifference – similar to that of the 'civilized world' as it watched this tragedy, this manifestation of utmost evil unfold. Nevertheless, there is one difference: in the civilized parts of the world, people today are not patting themselves on the back and seeking recognition as rescuers. Instead, they are building Holocaust Museums, erecting monuments of commemoration at sites where the Nazis perpetrated mass murders, and persecuting those who disseminate publications containing antisemitic diatribes. There are worthy role models in this regard."[1]

I have taken the expression 'the Classified Period' from notes in the margins of many of the documents that were signed by Ribbentrop and Molotov in their meetings in September 1939. This was the second round of talks, which they had started the previous month, in August 1939. Now they met to complete their deliberations. In this article I shall discuss one of the documents that was signed in this second round of talks - a paper that has become my source of reference as I attempt to study the fate of the Polish and Jewish populations in the period referred to in the title. The Protocol - the official title of this document – was one of the classified agreements signed by the two parties on September 29, 1939, after they had reached agreement on the final text. The following is a translation of this document (the translation and the emphases are mine – M.M.).

Classified German-Soviet Protocol – Addenda to the 'German-Soviet Agreement on Borders and Friendship

Upon completion of the German-Soviet Pact on Borders and Friendship, attorneys for the undersigned reached the following mutual understanding:

> Neither party will tolerate in its territory any Polish propaganda that is liable to harm or undermine the territory of the other. [Both parties] will frustrate any attempt at such propaganda in their territories and will inform each other of the means taken towards that end.

On behalf of the Reich – von Ribbentrop,
In the name of the Soviet Government – V. Molotov.[2]

As aforesaid, the signatories declare that they discussed the issue after the Pact 'on borders and friendship' had been signed, and reached 'mutual understanding' on the question of the policy to be adopted in the occupied parts of Poland. Their shared purpose was to implement a policy of imposing complete silence and a strict ban on the imparting of any information that could reveal developments in areas under the control of the occupying forces. In fact the real purpose of these talks was to impose a total blackout on the humanitarian conditions and the fate of both the Polish and the Jewish civilian population.[3] In other words, the two partners to the pact discussed what means and procedures they would adopt in order to prevent dissemination of any information, clamp down on any protest against persecution and killings, and ensure that international public opinion remained unaware of their plans for Poland and its population. Where the Germans were concerned, this referred to their arbitrary plans to uproot and transfer the population of ethnic Poles (Polish and Jewish alike) living in the territory that was to come under German rule, a plan involving requisitioning of property and above all, negation of legal status.[4]

As for the Soviets – this referred to their intentions regarding the Poles who lived in those eastern sections of Poland that had been annexed by Russia, to uproot them and transfer them to Siberia. Their plans also related to the Polish army personnel who had surrendered to the Soviet forces when the latter invaded eastern Poland.

The Protocol, and the secret intentions which it promoted, were thus an integral part of the policy of brutally silencing the Poles so as to facilitate the disintegration and division of Poland as a nation within the legitimate establishment of Europe. Hitler said this explicitly.[5] The Soviet Union too, as we shall go on to show, adopted a similar policy.

Furthermore, it was actively involved in the implementation of this policy, as demonstrated by the murder of the Polish officers who were illegally imprisoned by the USSR as prisoners-of-war in April 1940 at Katyn;[6] but not by this alone.

The Protocol, as aforesaid, had the most terrifying implications for Polish Jewry. Both parties were careful not to refer explicitly to Jews or acknowledge their having any recognized political status whatsoever, particularly in the circumstances created by the occupation. The Jews were thus left in a legal vacuum. In the Nazi-occupied territory of Poland, the Jews were deprived of all civic rights and were left abandoned and defenseless to face the occupying power. As for the Germans, this situation did not seem at all remarkable to them, since the 'Nuremberg Laws' had long ago determined that German Jews were sub-human. That, too, is how the Germans related to the Jews of Austria with the Anschluss of April 1938, and to the Jews of Czechoslovakia after it became a German protectorate in March 1939.

Just what the Germans had in mind for Polish Jews was revealed by Reinhardt Heydrich when he addressed commanders of the 'Einsatzgruppen' on September 21, 1939 in Berlin, in a message that came directly from Hitler. It was worded as follows: "The removal of the Jews to a foreign-language zone (Fremdsprachiger Gau) and their transportation to beyond the Demarcation line have been approved by the Führer, however, planning should be aimed at completion of the process within one year."[7]

In my article "Was the Jewish Question dealt with in the Molotov-Ribbentrop Negotiations?"[8] I discussed the hidden logistic reasoning behind Hitler's instructions. I maintained there that on September 21, the German leadership regarded the Visla river as the demarcation line between the areas of German and Soviet rule in occupied Poland. This division was discussed and determined in talks held in August 1939 between the two parties, based on the assumption that Germany would derive benefit from the establishment of what it viewed as 'the State of the remnants of Poland' that would be strictly limited to the Polish ethnic provinces. 'This State of the Remnants' was intended to encompass the provinces of Krakow, Radom, Kielce and zones in the eastern part of the Warsaw District, including the city of Warsaw itself.

The provinces east of the Vistula were decreed to come under Soviet rule, while those to the west were to come under German domination. In other words, what was envisaged was a condominium of the two occupying powers. It is this concept that gave rise to the Nazis' hope that they would be able to exploit this new set-up so as to be able, within one year, to complete the transfer of all the Jews in their areas of occupation to the Lublin Province and to the districts east of the Vistula within the Province of Warsaw – provinces that, as aforesaid, were part of the 'State of the Remnants' and now came under Soviet control.

However, this scheme never came to fruition. It was overruled by the Soviet leadership, which changed course on all matters pertaining to the idea of a reduced Polish state. Thus, in the second round of the Molotov-Ribbentrop talks in September 1939, the Soviet Union put forward new proposals that negated the entire concept of such a state. The Soviets proposed transferring the entire province of Lublin to Germany, in exchange for which Lithuania would come under their control, as had been the case earlier with Latvia and Estonia, whose fate was determined in the August round of talks. In light of these changes, the Germans at this stage of the negotiations proposed an exchange of populations that would include the Jews in reciprocal transfer arrangements. They went on to suggest that Birobidzhan and the Western Ukraine would be best suited for such transfers.

This came to light in a letter from Yegevni Chekmenev, the high-ranking Soviet official at the head of the Committee for Population

Transfer that was set up under another classified agreement between Molotov and Ribbentrop during their talks in September 1939 (which we will discuss below in greater detail).

However, the Soviet attitude to the Jews who remained in the Nazi-occupied territory of Poland was of itself both astonishing and terrifying, raising questions which we will consider in greater depth.

In the course of the occupation of Poland and its satellites, Polish Jewry could be divided into three categories: the first – the Jewish inhabitants of Polish territory annexed by the Soviet Union; the second – the Jewish inhabitants of Polish territory which was partly annexed by Germany and partly brought under close and exclusive German control; and the third – all the Jewish refugees who had fled the German-occupied areas to seek shelter in Soviet territory.

With regard to the first of these categories – the Jews in Soviet territory: there is no mention of them or their ethnic characterization in any of the documents signed between the parties. In other words, as far as the signatories to these documents were concerned, they had no political status whatsoever.[9] In the German-occupied territory they were ostensibly banded together with the Poles, and whatever applied to the Poles was to apply to them as well. There is no hint in any sources that the Soviets ever made any suggestion to the Germans - at the diplomatic or any other level - that these Jews be defined as a separate ethnic group. In any event, there is no mention of any legal rights they might have or expression of concern over the violation of their legal status.

We can learn what the Soviet attitude on this question was from records of the talks that Molotov held with Schulenburg, the German ambassador to Moscow, prior to the Soviet invasion of Poland in the period between the 1st to the 17th of September 1939, the day the Soviets crossed into Polish territory. The Soviet leadership was pre-occupied with the question of how to explain this surprise move to the outside world – a move not hinted at in any way whatsoever in the official statement on the Non-Aggression Pact between Germany and Russia of August 23. On September 10, Molotov informed the German ambassador to Moscow that the Soviet Union intended to issue the following statement to explain its invasion of Poland:

> The Soviet Government, basing itself upon the assumption that the German forces will continue to advance, has concluded that Poland as a state is disintegrated and consequently the Soviet Union finds itself obliged to come to the aid of the Ukrainians and the people of Belarus **who are under threat from the Germans** (the emphasis is mine – M.M.).

Molotov also asked the Germans to display understanding for this formula, which absolved the Soviets from being viewed as the aggressor in public opinion. But the Germans firmly rejected this request. On Sept. 16, Molotov presented a new formula to the German ambassador:

> The Soviet Union feels itself obligated to enter [Poland] for the sake of the security of its Ukrainian and Belarusian **brethren** (the emphasis is mine M.M.), and to ensure possibilities of peaceful work for this oppressed population.[10]

This time Molotov appealed to the Germans to approve this text. Both texts indicate that Moscow had decided not to include the Jews among the 'oppressed population' who were to be taken out of danger. After all, it hardly seems likely that the Russians were unaware of the dangers that threatened Polish Jewry from the Nazi occupation. Nor should we assume that they were ignorant of the manner in which the Germans intended to deal with the Jewish population of any territory that came under their exclusive rule.[11]

Even after this, the Soviet administration in the Polish territories annexed by the Soviet Union did not accord any specific legal status to the Jews as a group with special needs of its own. Formally, in this context, they were treated as Ukrainians in the Ukraine and as Belarusians in Belarusia – a formula that did not come about by chance. In other words, while Soviet policy towards the Jews during the 'Classified Period' might ostensibly appear astonishing, to be considered a passing episode in the circumstances of the time, looking at the situation retroactively we can only conclude that that was far from the case. This policy, as I will attempt to show, was a direct continuation of previous Soviet policy towards the Jews in Soviet territory and it can be understood in no other way.

A retrospective glance

In the context of the purges conducted in Soviet Russia in 1937-1938, the entire leadership of the 'Yevsektsia' was eliminated: all the Jewish posts on the Commissariat on National Affairs were annulled, as was the competence of its representatives on the Supreme Council of Government in the Republics and the Supreme Council of the USSR. Almost all the Jewish Party members of Lenin's generation and of the generation that joined the party during the Revolution and the Civil War were charged with various 'offences' and shot in the dead of night.[12] That is how Stalin dealt with the leadership of the Jewish national movement which emerged in the Soviet state after the Revolution, eliminating any hint of Jewish national aspirations that had begun to emerge.

The Yevsektsia was an organization whose ideological roots and political aspirations dated back to the Jewish national awakening of the nineteenth century in eastern Europe, an organization that, in its consolidation and struggle for its rights and authority, fought to reflect the historical continuity of Jewish national aspirations. The Yevsektsia was of course dependent on the powers invested in it by the regime, but tried nevertheless to abide by its political aims, which were an integral part of the national aspirations of the Jewish people. It collapsed under Stalin's murderous dictatorship.

Prominent among those assassinated were representatives of the Jewish workers' parties: the Bund, Poalei Zion, the Jewish Territorialist Workers' Party, the Jewish Autonomist Labor Party (the 'Seimists') and 'Zeirei Zion' (Zionist Youth).

The Jewish supporters of the Bolshevist party were motivated by two major ideas at that time: first, the belief that the October revolution heralded the great reform encapsulated in the socialist idea; and secondly – and this was the factor that convinced them – that the Soviet regime could not possibly ignore the Jewish question and would be forced to lend a hand towards the territorialization of the Jewish people, who would form a sovereign nation within the Soviet Union. We know that Lenin originated this idea back in 1913,[13] and went on to develop it in discussions he held with Simon Dimenstein[14] in January 1918. It was here that the basic ideas were formulated for the Commissariat on Jewish National Affairs.

Already by the early 'twenties, the idea was circulated of re-directing 100,000 Jewish families to agriculture in a so-called 'compact territory' that could form the basis for a Jewish Soviet Socialist Republic in the framework of the USSR. In geographical terms, the idea centered on the southern part of the Ukraine, close to the Jewish 'Pale of Settlement' that had been planned, and was in fact launched back in the days of Alexander the First; and in North Crimea. All the plans were directed at an area close to the Sea of Azov. It was with this project in mind that the Soviet authorities in 1924 established the 'Government Authority for the Transfer of the Jewish workers to the land' (the Komzet) as well as the 'Society for the encouragement of Transfer of Jewish laborers to the land' (Ozet). Among the functionaries of the latter were leading Jewish public personalities – some of them party members, others non-party. It is important to recall this chapter from past history in the context of my contention that, during the 'twenties', the Jews of the USSR were recognized as a national group with its own national objectives, aims that the Soviet regime both accepted and encouraged. The Komzet also received significant assistance from the Jewish associations – the Joint, ICA and ORT.

This is not the place to deal with the roundabout way in which this Authority went about furthering its aims.[15] Suffice it to say that the entire enterprise was doomed with the Birobidzhan plan of 1928. This area was given but scant mention in the plan put forward by Willem Lazis, a member of the college of the Commissariat for Agricultural Affairs of the Federative Republic of Russia and representative of the Komzet executive.[16] The fact that he put forward his proposal at the fourth session of the Komzet executive on January 8 1925 revealed the difficulties encountered in implementing the Komzet's plan for the re-settlement of Jews in the southern Ukraine and Northern Crimea.[17] Lazis' program, in fact, would have resulted in the dispersal of Jewish settlement to vacant areas of land in the Russian Federative Republic that were separated from each other by a distance of hundreds and thousands of kilometers.. One of these was a negligible part of what in the future would be known as Biro-Bidjan. According to this scheme, the total number of Jewish settlers who were to be absorbed in these widely scattered areas of land was not to exceed one hundred thousand Jewish families, the number specified in the Komzet document. This scheme for the dispersal of the settlers in fact put an end to all hopes for the eventual establishment of a Jewish Soviet Republic in the Soviet Union. It is thus hardly surprising that it was met with opposition in the Komzet executive and in the Ozet. Opposition to the scheme was led by Yuri Larin.[18]

The open and hidden confrontation went on for two years. Opponents of the scheme were unable to thwart the intentions of the highest echelons of government. Under its last initiative, a research delegation was sent to Birobidzhan and presented its findings after a short stay: it came out in favor of Birobidzhan as a suitable area for Jewish re-settlement. Objections to this scheme proved of no avail; the Komzet executive was 'persuaded' to issue a statement on January 17 1928, officially asking the USSR's Committee on Population Transfer to designate the area of Birobidzhan, which was named Birobidzhan Rayan – a new administrative district – to meet the requirements of Jewish re-settlement. The executive also instructed the secretariat of the Authority to go ahead with practical preparations without further delay. In an official notice dated March 28 1928, the presidium of the Central Committee of the USSR issued a positive response. Of special significance in this statement is its fifth clause: "Once satisfactory results are achieved in a compact area of the region in the re-settlement of Jewish workers, the possibility will have to be considered of this province becoming a **Jewish national administrative** and **territorial unit** (my emphasis – M.M.)."[19]

When this 'reassuring' response was given, there wasn't even a quorum of Jewish settlers in Birobidzhan. Four years later, on September 30 1931, when the Jewish population of the area had grown to 3,500, the all-Russian central executive announced its decision to define the area as a 'Jewish national unit'. On May 7, 1934, when the Jewish population had grown to 7,500 out of a total of 51,000, the Jewish district was declared to be an "autonomous Jewish province".[20] I mention these facts because they form the context in which we can study the paradoxical nature of these developments. During the early 'twenties, the Jewish leaders who hoped and strove for Jewish re-settlement in the southern Ukraine and northern Crimea could only dream of the Soviet authorities issuing a declaration of this nature in regard to these districts, fully aware of the weight and significance such a declaration would carry. But that was not to happen. The reason given for such evasions was that it was best to refrain from declarations of policy until such time as a large, compact Jewish settlement emerged in a specific territory.

More on Latzis' Scheme

When Willem Latzis proposed including the Soviet Far East in his scheme for Jewish re-settlement, he spoke of three small areas in that context. He dwelt on the objective difficulties of adapting the regions to agricultural settlement and the estimated number of families that could be re-settled in them. He outlined the borders of these areas, stressed the work on infra-structure that would have to be carried out before any settlers could be absorbed there, and estimated the number of families that could be settled in each area.[21] From Latzis' detailed plan we learn that he foresaw the re-settlement of a total of 15,000 families in three separate areas, while a further 85,000 were to be dispersed – in the spirit of his scheme – among settlement blocs that were hundreds of thousands of kilometers apart from each other. Therein lay the weakness of his proposal: this factor made it unpalatable to the Komzet.

And it is here that events took an unexpected turn, a development originating with the Politburo: as the Komzet was grappling with this difficult situation, it proposed a new solution. It took a neglected area, the most difficult in Latzis' entire scheme for the Far East – the area between the Bira and the Amor rivers – and expanded it through the annexation of other areas of the same nature bordering it to a total area of 37,000 sq. km, declaring this to be the answer to the dream of a Jewish republic. However, this was a mixed blessing: tremendous investments would be needed for adaptation of the terrain and soil amelioration. Furthermore, even if it were scientifically and technically possible to carry out the plan, a tremendous input of voluntary endeavor would be required. Despite the waves of enthusiasm and support that swept through Jewish communities abroad on learning of this declaration, neither the billions that would be needed for the scheme nor the requisite pioneering manpower were forthcoming. For despite strenuous lobbying, the scheme won little support among the Jews of Russia, who harbored few illusions about the nature of Birobidzhan.

As far as Stalin was concerned, he viewed the Birobidzhan scheme as giving the Jews the opportunity to 'play at a Republic'. The two to three years that the district functioned under a Jewish administration were quite enough for him. In 1936, the NKVD 'discovered' that Joseph Lieberberg, chairman of the district executive, a man of many talents and director of the Jewish Scientific Institute in Kiev, who had volunteered with his entire staff to move to Birobidzhan to contribute to the cultural reinforcement of the new Jewish district – was a former member of the 'Trotskyist' Poalei-Zion movement. This 'discovery' determined his fate – a fate similar to that which loomed over the entire leadership of the district. In 1938, Birobidzhan became a fiction of a Jewish province, which is how it continues to exist to this very day. Why is it maintained to this day, what is behind this? Perhaps Putin has the answer.

The 'purge' of the Jewish leadership in the Soviet Union in general, and in Birobidzhan in particular, was not specifically directed against the Jews. It was part of the great purge in which, it will be recalled, Russian Communist Party members from several generations were eliminated, as were also the heads of national leaderships of various republics of the USSR as well as of the autonomous republics. After the Great Purge, the Federation of Soviet Republics itself became a fiction – and Stalin saw his dream come true. Immediately after the Bolshevik victory, he began to regard the Soviet national republics as a danger to the unity of the USSR and defined the trend to promote them as 'that dangerous game of republics'.[22]

After the Great Purge, a vigorous process of Russification was set in motion and the fiction of the Soviet nation was vigorously promoted. The Jews were among those crushed under the wheels of this process.

From this point on, the Jewish educational system began to dwindle and Jewish institutes of research simply disappeared. Lone remnants of a flourishing Jewish culture were the Yiddish theater and the Yiddish writers' sections under the Writers' Associations of the individual Republics. Even so, there remained a hidden but most interesting Jewish presence: I refer to the Population Census, in which the nationality of citizens was recorded according to their genetic origin, as well as any possible changes due to mixed marriage. All these details were diligently recorded in the files, as they might later prove useful to the Internal Secret Services (the NKVD). In other words, from this point on the Jews continued to exist in that grey 'formal' annotation that, while it gave them no ethnic rights, answered to a kind of folkloristic nostalgia and to the intrigues of the ruling authorities.

It is therefore hardly surprising that, upon the annexation of eastern Poland to the Soviet Union, a similar method was applied there as well in regard to the requirements of the Jewish population. No public declarations of principle were issued, the government did not lay down any rules, but only responded to the wishes of the local residents. As for Hebrew education: schools where Yiddish was the language of instruction were allowed, and – where this was agreed – Jewish schools where Polish was the language of instruction were also permitted, mainly in the larger towns. However, already by the end of the first school year, the government initiated parents' meetings at which parents (guided from behind the scenes) posed loaded question such as whether Yiddish-language instruction would help the students secure places at colleges and universities in the Soviet Union, to which the inevitable response was that it would not, and that it would be better to switch from Yiddish to Russian.[23]

Jewish inhabitants of Polish territory were partly annexed by Germany, while the other part came under exclusive and firm Soviet control

The Soviets, totally indifferent, left the Jews to the fate prescribed for them by the Nazis. Thus for example, Jewish soldiers who had served in the Polish army and captured by the Soviets were returned to the area under German occupation. On October 21 1939, an agreement was signed for an exchange of prisoners from the Polish army who had been held in German and Soviet POW camps. The Germans undertook to return any prisoners whose permanent home address was in eastern Poland to the Soviet territory, while the Soviets were asked to release POWs originating from the parts of Poland that were now under German control to German territory. It soon became evident, however, that there were Jews among the POWs held in the Soviet zone who were far from overjoyed at this arrangement, perceiving the dangers involved. They asked to be spared this fate and appealed to the Soviet authorities to be allowed to remain in the USSR, some claiming that they had close family members in that area and that they were former Communists, while others based their claims on their keen support for the regime.

The latter explicitly spelled out their fear of being trapped in German-occupied territory. In response to these pleas, L. Beria on November 4 issued the following order to the authority in charge of repatriating prisoners: "After consideration of the requests of prisoners-of-war not to be sent to their (former) places of residence (in territory held by the Germans)… the Authority responsible for prisoner-of-war affairs finds the reasons given for these pleas to be unconvincing."[24]

We thus see by their reply that the Soviet authorities deliberately refrained from noting the ethnic identity of the POWs who had asked to remain in Soviet territory and callously abandoned them to their fate. This attitude is also reflected in the following letter of complaint that was presented to Schulenburg, the German ambassador to Moscow, by the deputy commissioner for foreign affairs, F. Potiemkin on December 17, 1939:

> The Foreign Affairs Commission has received information on a surging crowd of refugees, mainly Jews, who are determined to cross over to the Soviet side in their masses.
> 1. On November 10, near the town Leska south of Przemysl, 7 people crossed the border. One of them was shot by German forces.
> 2. On November 10, in the vicinity of Przemysl, 11,268 people crossed the border. Of these, 450 were sent back to Puhle from the German army HQ. When the Soviet party attempted to send back the rest, Feuer, an employee of the German border service which had agreed to take them back, opened fire on them and killed 20 out of the first 30 who were sent back.
> 3. On November 13, 16,500 people crossed the border, 3,5900 of them via Augustow, 5,000 of them via Lomza, 3,000 via Szepietowo and 5,000 via Brzesc.
> 4. On November 29, in the vicinity of Szepietowo, 2,500 crossed over. After they were sent back, they were all shot by the Germans.
> 5. On December 7, 2,500 crossed over via Sokal and Pruszyna. After they had been given due warning, they were released on the demand of the Soviet side by the commander of the German border services, Leitung.
> 6. On December 12, 2,000 crossed over but were sent back to the German side.

The letter of protest also included a demand; 'Let the Germans keep their Jews in their own midst'.[25]

The Jewish refugees from German-occupied Poland in Soviet territory

One of the most troubling problems, the handling of which revealed covert intentions, was that of the Jewish refugees from the German-occupied zone who sought refuge in Soviet territory. On October 1 1939, only thirteen days after the German invasion of Poland, the Soviet Communist Party's Politburo convened to discuss the problem of these refugees. Two of its distinguished members, Lavrenti Beria and Lev Machlis (a Jew), were assigned to present 'proposals on the question of prisoners-of-war and refugees' within three days[26]. We do not know whether the Politburo had already specified its intentions at this stage, but going by the usual practice, it seems most unlikely that no recommendations had been drawn up. Nor do we have any information on whether Beria and Machlis did indeed present their conclusions within three days. However, we do have a document of the Politburo dated November 10, 1939 on 'refugees from territory currently occupied by the Germans'. There were two clauses to the resolution passed after the discussion:

> 1. In order to find a solution to the problems relating to utilizing the labor force of the refugees, including the question of repatriation of some of them, it is decided to set up a committee composed of Beria, Nikolai Bulganin, Nikolai Schwernik and Leonid Konietz, Beria to be co-ordinator of the committee.
> 2. This committee is to be authorized to take an accurate census of the refugees and to **examine the repatriation of the rest of them.** The committee is to put its recommendations before the Central Committee for ratification.[27] [the emphasis is mine – M.M.]

The document tells us nothing about the fate of the Polish prisoners-of-war, and one can only surmise that a separate decision was taken on this issue. Perhaps we should look for it in the Soviet documents on the Katyn affair. In any case, we know that the POWs were placed under the supervision of the NKVD – a measure that was in violation of the rules pertaining to prisoners-of-war. As for the refugees, the resolution called for a census to determine exactly how many there were, and to consider the idea of repatriating those who could not be integrated in available work in Soviet Russia back to Germany. Such refugees could not remain in Russia and the Politburo was not prepared to consider other solutions for temporary arrangements in the USSR. It should be noted that the resolution does not refer to the national identity of the refugees given that, under Soviet doctrine, even in this context there was no such thing as a specific Jewish identity.

Indirectly, this could be taken as further proof of the basic indifference of the Soviet leadership to the fate of Polish Jewry, and of its desire to evade dealing with this problem (Stalin, it should be noted, was present at these sessions of the Politburo, which in turn would pass no resolution without his consent.) The Communist Party's Politburo might have decided to repatriate 58,000 refugees to Germany as early as December 1, 1939.[28]

From the findings of the census of refugees that was submitted to the Politburo, we learn that at the beginning of December 1939, there were 121,251 persons defined as refugees on Soviet territory. The committee on re-settlement offered employment to 63,250 of them, and suggested the rest be repatriated to the German side. The number of refugees confirmed by the Politburo in this resolution greatly differs from the numbers that were recorded in various studies of the period.[29]

We may reduce this difference if we assume that the large numbers previously reported in research related specifically to Jews who had been present in the west (or German zone) of Poland at the time of the invasion and had fled to the east. However, the refugees also included people born in eastern Poland who had moved to the west for better job opportunities in the circumstances that had arisen. Such people made every effort to return to their families and ensure that they had not lost their residents' rights, and consequently were not counted as refugees when the census was taken.

As for those refugees the Party's Politburo intended to repatriate to Germany: it emerges that already on December 20, the Politburo was forced to annul its decision of the 4th of the month, as the Germans refused to take back the Jewish refugees. It would appear to be shortly after this that the Politburo decided these refugees should be sent to the northern provinces of European Russia (the Kirovsk and Arkhangelsk provinces).[30]

Another classified protocol from the September talks between Molotov and Ribbentrop

In the context of the understandings reached and signed during the Molotov-Ribbentrop talks on September 28, the repatriation to Germany of the German population of eastern Poland and the Baltic countries was discussed in a separate category. The protocol of this agreement contained two clauses: the first presented the German wish that the Soviet Union allow the German citizens of the aforesaid areas, and the Germans who had settled in the areas in the recent or distant past but did not hold German citizenship – to return to Germany voluntarily. In this clause, the practical aspect of implementing such a transfer was also discussed: who would be the negotiators, as well as issues of property. The brief second clause conveyed the Soviet wishes: that the government of the German Reich undertakes a reciprocal commitment with regard to refugees **of Ukrainian or Belarusian origin** who are resident within its sphere of interest."[31] (the emphasis is mine – M.M.)

The Soviet side had a broader version regarding those who had the right to be transferred to Soviet territory, as revealed in a letter from Chekmeniev to Molotov dated February 12, 1940. Included among those who had such a right were Russians and 'Ruthens' (the name given to the Ukrainians by the Habsburg Empire in Galicia prior to the First World War).[32] However, another relevant point arises from Chekmaniev's letter: namely, that the Germans attempted to include the Jewish population among the subjects of this classified Protocol and, under its auspices, to carry out a transfer of the entire Jewish population of the areas under German control to Russia. The Russians did not categorically reject this notion, and the Germans apparently got the impression that the practical aspects of the plan would be determined at a later stage. This is what is behind the question in a letter they addressed to Chekhmaniev –"When will we start organizing transportation of the Jews to Birobidzhan and the Western Ukraine?" We may be fairly certain that the Russians answered the Germans in the negative. Nevertheless, one cannot escape the conclusion that this agreement, hidden in the text, was of astonishing significance.

The Russians, the Ukrainians and the Belarusians are genetically part and parcel of the Soviet Union; they are its brethren and it is committed to them, whereas the Jews are a foreign body, an alien race, and the Soviet Union has no obligation to take care of them. Of course some will argue that the text specifically refers to Jewish citizens of Poland, but that leads to the inevitable question – were not the Ukrainians and the Belarusians referred to here also citizens of Poland?

Furthermore, when the details of the agreement on population exchange between Germany and the Soviet Union were published in the

German-occupied zone of Poland, many Jews believed it provided them with the opportunity to flee to Soviet territory. The German authorities also welcomed this development and tried to help the Jews to 'infiltrate' the Ukrainian and Belarusian convoys advancing to the east. However, when these convoys reached the border, they were met by Soviet border officials who checked the ethnic identity of all the refugees, picked out the Jews and sent them back to where they came from.[33]

To conclude: the Politburo discussed the fate of the Jewish refugees, deliberately ignoring their ethnic affiliation, and decided they be expelled to Germany under circumstances in which they, like the rest of the world, were fully aware of the fate that awaited them in that country. The Politburo deliberately resolved not to publicly announce its decision, with the clear intention of concealing its criminal nature. It added insult to injury by expelling them – entire families with their young children and grandparents – to the areas of the toughest, wintry, icy conditions in northern Russia. Admittedly, the Jews were eventually saved through the agreement between the Soviet Union and the Polish Government-in-Exile led by General Wladislaw Sikorski (after June 1941), under which they were permitted to move to climatically less traumatic areas. Objectively, they were saved from the certain death faced by Jews in the German-occupied territories of Soviet Russia, but one can hardly say such mercies were born of goodwill.

Soviet Propaganda and the 'Classified Protocol'

It would be mistaken to assume that the classified protocol that was signed by Molotov and Ribbentrop was applicable only to the Polish question. It quickly emerged, to the astonishment of many, that this Protocol inevitably had implications for the entire system of research and propaganda that the Soviet Union could operate within its own territory, as well as regarding its efforts to deal with world public opinion – particularly in regard to the Comintern, which had to convey instructions to the many different communist parties that came under its authority.

Indications of the changes that were bound to take place in Moscow's attitude to Nazi Germany began to surface at the official ceremonies that accompanied Ribbentrop's presence in the Soviet capital for talks with Stalin and Molotov, who celebrated the new age in German-Russian relations by raising a toast to the German nation and its leader.

In addresses he gave on August 31 and October 31, Molotov stressed that it was essential to stop portraying Germany in a hostile manner and put an end to the anti-fascist propaganda. He declared that at this time the real warmongers were England and France, while Germany sought peace. Molotov argued that narrow-minded people had been misled by over-simplified anti-fascist propaganda and were incapable of understanding the improvement that had taken place in Soviet-German relations. He went on to maintain that the terms 'aggression' and 'aggressor' could no longer be applied as they had been previously, since Germany was now a peace-seeking nation, while France and England wanted the war to continue and the British government had actually declared its intention to 'eliminate Hitlerism'. Molotov said:

> One may support or negate the ideology of Hitlerism, that's a matter of opinion, but to seek to settle the issue by means of war – that is not to be. That would be not only an unwise course but **above all a crime** - and this is a war that is being waged under the false banner of democracy." [34] (the emphasis is mine – M.M.)

The Soviet authorities did not hesitate to enforce acceptance of the new situation, even on the cultural and literary activities of Soviet citizens.

Officials of the central authority for control of the repertoire of theatres, cinemas and published literature were required to re-examine 13,220 works of different categories of art, of which they banned 4,200. Plays that were running in the theater were cancelled or had their texts amended in such a way as to erase any hint of anti-fascist or anti-German sentiment. The movies, 'Professor Mameluke' and 'The Oppenheimer Family,' which expressed sympathy for the plight of the Jews, were banned from the screen. In Moscow's central library, readers were not allowed access to any anti-fascist newspapers; their places on the shelves were taken by Nazi publications from Germany.[35] These are but a few of the changes introduced as a result of this 'new thinking'.

The Soviet Union thus stopped all anti-Nazi propaganda. It is therefore hardly surprising that we do not find any articles about Hitler, German fascism or Nazi Germany either in Soviet newspapers or in historical, political or ideological periodicals of this period. The articles that do appear in these publications tend to condemn France and Britain, particularly their imperialism, claiming that German imperialism is young and far less dangerous, and that there is just cause to its territorial aspirations. Above all, they stressed - Hitler has not ceased to offer peace.

The Comintern in the service of Soviet politics, and the Jewish aspect

The Comintern found itself in an embarrassing situation, as did the Communist parties that struggled to find direction in this new situation. After the first agreement of August 1939, they tried to explain these developments as a great achievement for Soviet diplomacy that was in fact restraining Hitler and soothing international tensions. These reactions served to calm down the Comintern, so long as the details of the secret agreement between Germany and the Soviet Union on the division of Poland and the fate of the Baltic countries were not known.

When the Germans launched their invasion of Poland at the beginning of September, G. Dimitrov, head of the Comintern, and its secretary Dimitri Manuilski approached Stalin about the guidelines to be issued to the Communist parties. On September 5, Dimitrov sent Andrey Zhdanov the draft text of a notice he intended to circulate among the parties. Even before it was published he noted:

> As we spell out instructions for the course of action, and in particular the tactical positions and the political aims of the communist parties in the new circumstances we find ourselves facing extraordinary difficulties. In order to overcome them … we are more than ever in need of the direct assistance and advice of Comrade Stalin.[36]

And indeed, on September 7, a meeting with Stalin took place. Dimitrov made the following notes of Stalin's explanations in his diary:

> **Stalin:** The war is being fought out between two groups of capitalist countries (the wealthy and the poor in terms of colonies, raw materials etc.) with a view to re-dividing the world, with each seeking to acquire more power. We are not opposed to their fighting it out between them and thereby weakening each other. It would not be a bad thing if Germany were to bring the wealthy capitalist countries (particularly England) to their knees. Hitler himself does not understand – and indeed this is not what he is after – that what he is doing is undermining the entire capitalist system. The position taken by the communists where they are in power differs from that of the party where they are in opposition. In our own house it is we who are in charge, whereas the communists in the capitalist countries are in opposition. There the bourgeoisie is in power. We can manipulate, urge on one side in its battle

> against the other and stimulate the fight. The non-aggression pact may to some extent be of help to the Germans. Now the next stage towards achieving our purpose will be to encourage the other side. But in the capitalist countries, the communists should be vigorously opposing their own governments and coming out against the war.
>
> Until the war broke out, it was **absolutely correct** to support the confrontation between the democratic form of government and fascism, but once the war between the imperialist powers broke out, that is no longer the case. The division of the capitalist countries into fascist and democratic states no longer has the same significance as it did before...
>
> Earlier (in the course of history) the Polish state was a national state and as such it was defended against division and suppression by the revolutionary forces. Today it is a fascist state that oppresses the Ukrainians and Belarusians and others. In current circumstances, the significance of the elimination of that state would be that there is **one fascist bourgeois state less**. What would be wrong with (a situation) where, as a result of Poland's defeat, we were able to disseminate the socialist system in new territory and among a new population? [37] (the emphases is mine – M.M.).

According to Stalin, in the new situation, the division of the capitalist countries into fascist and democratic states no longer had the significance that it had in the past. It did not take long for Molotov to adopt this line: in an address on October 31, he labeled England as 'criminal' for declaring that its purpose in declaring war on Germany was to overcome Hitlerism. Molotov said that one's attitude to Hitlerism was a matter of opinion, and that war was certainly not the right way to uproot it. And yet fascism surfaced again when Stalin referred to Poland, which he described as a fascist state, adding that the country deserved its fate.

In any case, he added, there was some benefit to the fall of Poland – now there was one fascist state less. These remarks were most embarrassing to all the communist parties that should have been fighting against Nazi occupation in their respective countries – the Netherlands, Belgium and France in 1940 and Greece and Yugoslavia in 1941, and above all to the German communists.[38]

To this we may add the Jewish aspect – the situation and fate of the Jews under Nazi rule. Thus we find that, according to Stalin, fascism was not such a bad thing in itself; it was not a cause for existential conflict and was simply not a particularly significant issue. Fascist ideology was of no particular concern (to Stalin) and even Hitler's extreme, murderous hatred of the Jews was not a cause for urgent confrontation. This issue was merely one of outlook and therefore the question of the fate of the Jews should not be of particular concern to the 'socialist' nation.

Molotov took up this theme, using remarkably mild language in discussing the range of ideological differences that the Soviet Union could tolerate. In other words, it was no crime to abandon the Jews to their suffering and their cruel fate. It was reasonable behavior and, as I see it, to Stalin it was not undesirable.[39]

The collection of documents from the Comintern archives that was published in English by Yale University contains letters and drafts for instructions issued by the Comintern secretariat and by Dimitrov himself – drafts sent to Stalin in 1934-1935.[40] On September 7, Dimitrov met Stalin to discuss the line of propaganda to be taken after the outbreak of war. On this occasion, Stalin discussed the subject of Poland, saying that "it is a fascist country that oppresses the Ukrainians, the Belarusians and others." After further discussions, Dimitrov and the Comintern secretariat, under his guidance, drew up draft guidelines for evaluation of the situation that were sent out to the communist parties.

The guidelines were ready by September 26. Of special interest is clause 4 of this document, which dealt with Poland. Even though it is far from complimentary about Poland, this clause does not include any assertion that Poland was a fascist country. However, one paragraph dealing with the case in point stands out in particular:

> "This is not the Poland of 1848, which fought for its independence from Tsarist Russia. This is a prison of nations that has come to replace Tsarism and that oppresses the Ukrainians, **Jews**, Lithuanians and other nationals [under its rule]."[41] (the emphasis is mine – M.M.)

I assume that it was no coincidence that Dimitrov enlarged upon Stalin's "and others". Stalin's omission of the Jews appears to have astonished even the secretary of the Comintern and he felt that the organization could not possibly overlook the plight of the Jews. However, the draft that he prepared was not approved and no alternative text was ever drawn up. There was no further mention of the Jews even in Dimitrov's letters to Stalin (I assume that this was at the latter's request).

In 1940, the German army went ahead with its drive across Europe. As the German forces advanced, European countries collapsed one after the other – Denmark, then Norway and after that Holland, Belgium, France and later, in 1941, Greece and Yugoslavia. In each of these countries there was a Jewish population with which the Nazi occupying power intended to deal in its known fashion. However, there were also communist parties, the most influential of which was the French communist party, led by a large cadre of active members and with an impressive following of sympathizers, among whom there was a significant Jewish presence. During the inter-war years, France had absorbed a huge number of Jewish immigrants from Poland, some of whom who had come to look for work and others who had fled Poland for political reasons. Most of the latter were communists who were persecuted in Poland and sought refuge in France.

In 1940, the Jewish immigrant population in France was estimated at 150,000. The Vichy government was prepared to abandon these Jews to their fate at the hands of the Nazis, and ordered the French gendarmerie to round them up and hand them over to the Gestapo.[42]

There is no doubt that the French communist party was aware of the Nazi oppression of the Jews in France and of the agreements that had been entered into between the Vichy government and the Nazi regime of occupation. There are no grounds whatsoever for any suggestion that the Party took a positive view of these agreements, which leads to the question - did the Party know to what destination the Jews of France were being transported after they had been rounded up and herded into closely-guarded camps? What did the Party do to warn the Jews of the grave dangers facing them? It is my belief that the Party was unaware of the death camps and thus could not of course have given any clear or explicit warning.

During the 'classified period' from June 1940 (the surrender of France) to June 1941 (the German invasion of Russia), the French communist party followed the guidelines issued to all the parties by Stalin and the Comintern. They adopted the Soviet argument that the war had broken out due to conflicting interests between rival imperialist powers and that it was not a confrontation between democracy and fascism or, as Dimitrov phrased it in his guidelines "it is a bourgeois myth that the war is of an anti-fascist character".[43] Yet by that time, there were no longer any doubts about the nature of fascism – neither about its ideology nor about its practice of racism or its hatred of the Jews.

The French communist party was required by Stalin and the Comintern to foster loyalty to the German occupation - as can be seen by the fact that the possibility was discussed of *L'Humanité*[44] continuing to appear as a legal publication. Jacques Duclos, one of the party leaders who had remained in France and gone underground, reported to Moscow on the situation in France and the public mood. He also suggested the course of action the party should take in this complex situation. Yet in none of his reports was there any mention of the plight of the Jews in France or of how the party could help them. The Jews were simply not mentioned. And the same applies, of course, to Dimitrov's replies to Duclos.

The following spring, in April 1941, the Vichy police began to speed up the pace of the round-up of stateless Jews and of all those who had fought in Spain and sought refuge in France until the end of the hostilities there.[45]

When Germany began the invasion of the Soviet Union and the communist parties were released from having to abide by the Stalinist doctrine of the 'classified age', one might have expected that the attitude of the Comintern to the plight of the Jews would change, but that was not the case. We have found no documentation whatsoever on this matter in the Soviet documents that have so far been published. However, there is indirect evidence leading to a clear-cut and tragic conclusion. We know, for example, of an internment camp for communist Jews and former participants in the Spanish civil war at Beaune-la-Rolande in the Loiret district. There was an active communist underground at this camp, and the prisoners were led by a committee of communists and communist sympathizers who maintained close contact with the party in Paris. In April 1942, word came from the Party in Paris that as many Jews as possible should be encouraged to escape from the camp. In the second half of June 1942, the Germans began to transport prisoners from the camp over the French border. The Party sent instructions to members of the committee as to how the communists and sympathizers should act wherever they were sent, in the assumption that the prisoners were being dispatched to Germany to work in factories. In this context, the party instructed them:

1. To carry out sabotage; 2. Not to accept any responsibility or any function in this connection; 3. To fake illness whenever possible; 4. Not to give away any information on professional subjects, particularly anything that might be useful for military production.[46]

At the end of their journey, the prisoners from Beaune-la-Rolande were dismayed to find themselves in Poland – they were in Auschwitz. It is also worthy of note that the first train-loads of Jews of Polish origin who had been rounded up in Paris arrived at Auschwitz on March 27, 1942.[47] No reports on their fate were leaked to France.

This is astonishing, as such information could have reached the French communist party through its channels of communication with the Comintern and the Soviet communist party. Moscow broadcast over a large number of radio stations, yet apparently it chose not to give the rest of the world - particularly the communist parties that came under its auspices - detailed and accurate information about the fate of the Jews.

We thus see that, until June 1941, the policy laid down in the classified Protocol was observed. After that date the Soviet government, under Stalin's personal guidance, met the extermination of millions of Jews with icy indifference.[48] Indeed, it is a matter for profound speculation that the Soviet regime, which in actual fact abandoned the Jews to their fate, is regarded by many Jewish communities as the power that saved them. The rescue that originated out of this shameful omission has somehow come to be perceived as a great and altruistic act of sacrifice.

Appendix: A letter that Y. Chakmeniev addressed to Molotov on 9 Feb. 1940 Ref: 3440 ...

Text of the letter: USSR
To the director for population transfer of the Popular Comintern Council of the Soviet Union - Feb 9 1940.
No. 01471c [49]
Moscow, Krasniaya Ploshechskadez 3
To the Chairman of the Popular Council of the Comintern
To Comrade W. M. Molotov
In the Population Transfer Administration of the USSR [Council of the Popular Commissars of the Soviet Union]
Two letters have been received from the Population Transfer Bureaux in Berlin and Vienna on the question of organizing [....] the transfer of the Jewish population from Germany to the Soviet Union - in practical terms, to Birobidzhan and the western Ukraine.[50]...
In accordance with the agreement between the governments of the Soviet Union and Germany on transfer of population, the following are to be evacuated into the territory of the Soviet Union: Ukrainians, Belarusians, Rotanys and Russians.
In our view the proposals of the Transfer Bureau are unacceptable. Please give us instructions.
Appendix: 6 pages.[51]
CHK
Head of the Population Transfer Administration of the Council of People's Commissars of the USSR.

Notes

1. The title of the article in the periodical *Niezavisimaya Gazeta*: "The Soviet leadership looked on in ice-cold indifference at the fate of an entire people" ("Sovietskoie rukovodstvo c ledzianym bezrazliciem nabliudalo za sudzboi celovo naroda").
2. Max Domarus, *Hitler's Reden und Proklamationen: 1932-1945* (Wiesbaden, 1973). 1373. Hereinafter, Domarus.
3. See: Adolf Eichmann's notes on Heydrich's meeting with Einsatzgruppen commanders, Berlin, 21.9.1939 .The translation in the Hebrew edition of *Yalkut Moreshet*, 89.
4. Ibid., ibid.
5. Jochen Boehler. *Auftakt zum Vernichtungskrieg. Die Wehrmacht in Polen 1939* (Frankfurt a/M, 2006) 31.
6. Yakovlev, (Ed.) *Katyn, Prisoners of a War Never Declared* (Moscow, 1999). From the Preface: "The Allies (the Germans and the Soviets) made a reciprocal commitment to wage war against all manifestations of opposition in occupied Polish territory, and of course to inform each other of the measures they intend to take for this purpose". The editor concludes that "the possibility cannot therefore be categorically ruled out that the massacre in April-May 1940 of almost 22,000 Poles who were in camps and prisons in the territory of the Soviet Union, was coordinated with the Hitlerist 'Akzia AB' which was perpetrated at the same time (in the territory of the General-Government)", 16.
7. The foreign-language speaking Gau (district) is the area of western Poland excluding the districts of Pomerania, Posen and Upper Silesia which the Germans intended to annex without delay. At this stage of the talks between Germany and the Soviet Union, i.e. in the August meetings between Molotov and Ribbentrop, both parties considered the possibility of establishing a 'State of the remnants of Poland' the eastern part of which would come under Soviet and its western part under German control. The Vistula River was to have been the line of separation between the two parts. Thus what Hitler had in mind with his message referring to "transportation across the line of separation" was the concentration of all the Jews in the Lublin district, which was to become a Jewish 'reservation'. In the September talks between Molotov and Ribbentrop, the scheme for the 'remnants of Poland' was cancelled and new destinations were sought for the notion of transporting the Jews. *See* "Adolf Eichmann's notes on Heydrich's meeting with Einsatzgruppen commanders in Berlin – 21.9.1939", in this issue.
8. *Galed*, 22, 2010, 107-130.
9. In Hitler's view, the Jews were a clearly-defined group which should be deprived of all their rights and uprooted from their places of residence, whereas Stalin, in

discussing the Jews, concluded that "what we have here is not a living, functioning nation but something mystical, incomprehensible, unreal". Joseph Stalin, *On the National and Colonial Question*. Russian original: "*Poluchaietsia, ne nacia, zivaia, dzeistvuiuszchaia, a nechto misticheskoie, neulovimoie i zagrobnoie"* I. Stalin *Sochinenia,* t. 2 (Moscow, 1951), 301.

10. Slawomir Debski, *Miedzy Berlinem a Moskwa, stosunki niemiecko-sowieckie 1939-1940* (Warsaw, 2007), 115, 122. Hereinafter: Debski.
11. *See* my article "Was the Jewish question discussed in the Molotov-Ribbentrop; talks?" *Galed* 22 (2010), 107-129.
12. For the first detailed account of this outbreak see: Gennadi Kostryrchenko *Tainaia Politika Stalina, vlastch I Antisemitism* (Moscow, 2001), 123. (*Stalin's Secret Policy – Power and Antisemitism).* For some of the concrete evidence: G. Yakovalev (Ed.) *Rehabilitation* (Moscow, 2000). It includes a collection of documents from the Division. L.58-62, d. 389 opis 8, f. 3 RGANI.
13. See*:* my article "Lenin's hidden formula on the Jewish question and its presence in the Soviet-Jewish dialogue", *Iyunim, Studies in the Revival of Israel,* 16 (2006), 533-547.
14. Simeon Dimenstein – attended the Volozhin *yeshiva*; an erstwhile member of the 'Bund' and later of the Bolshevik party, he was an associate of Lenin who discussed Jewish issues with him. Lenin appointed him head of the Commissariat on Jewish National Affairs. A member of the Collegium of the Commissariat on National Affairs when it was under the chairmanship of J. Stalin, he acted as Stalin's deputy and was involved in important decisions on Jewish issues. One of the founders of the Komzet and a member of its executive. Murdered in 1938.
15. See: M. Mintz "The actions and confrontations of the Soviet establishment in promoting settlement of the Jews on the Land - 1924-1925*" Iyunim, Studies in the Revival of Israel* 19 (2009), 463-489, Herein after: Mintz, 2009.
16. Villem Latzis. After the October Revolution, he was a member of the Collegium of the Agricultural Affairs Commissariat and its representative in the Komzet directory. Executed in 1938.
17. See: my article on "Problems of the agrarianization and territorialization of the Jews in the twenties – according to the documents that have come to light in Soviet archives", *Iyunim, Studies in the Revival of Israel* 5 (1995), 167 ff..
18. Yuri Larin (Lurie), son of the officially appointed rabbi of Kiev. An economist; formerly belonged to the Russian Social-Democratic Party and supported the Mensheviks. He was close to Lenin and in 1917, joined the Bolsheviks. He supported the Komzet's initiative and was head of Ozet. Played an active part in the 1st conference of Ozet (November 1926). His book *The Jews and Anti-Semitism in the Soviet Union* was published in 1929 (in Russian).

19. Yaakov Livavi (Babitzki) *Jewish Settlement in Birobidzhan* (Jerusalem, 1965), 152-351 [Hebrew].
20. Ibid. 58.
21. Garf, f. 7541, opis I, d. 26 str. 21-23.
22. A. N. Yakovlcv (Sostavitiel), *Stalinskii plan po unichtozeniu naroda,* podgotovka I realizacia prikaza #0047 operazia po repressirovaniu byvschich kulakov, ugolovnikov I drugich elementov (Moscow, 2010) (A. N. Yakovlev (ed.) *Stalin's Plan for the Elimination of a People*. Emphasis is placed in it on the assassination of party members both in the central government establishment and in the autonomous socialist republics. On "The play in the republics" see: Mintz, in an essay published in *Iyunim, Studies in the Revival of Israel* 19/2009, 468-470.
23. Dov Levin, *A Period in Parenthesis 1939-1941* (Tel Aviv, 1989) 105 [Hebrew].
24. Slawomir Debski: "Niech swoich Zydow Nienmcy trzymaja u siebie" Zydzi w stosunkach sowiecko-niemieckich 1939-1940." *Arcana* 28, 1999. 56 1 d. (Slawomir Debski : "Let the Germans keep their Jews in their own midst").
25. Ibid., 55-56.
26. RGASPI f. 17, opis 3, d/ 1014, 1.47, 61
27. RGASPI f/17 opis 3. d. 1016, 1.3.
28. Debski, 189. RGASPI, f. 17, opis 3 d., 1018.
29. For a higher estimate of the number of refugees – see: Joseph Litvak: *Jewish Refugees from Poland in the Soviet Union 1939-1946* (Tel Aviv, 1968). Appendix from 353 [Hebrew].
30. Resolution of the Central Committee: 'On the Refugees', RGASPI f. 17, opis 3, d. 1018, 1. 4.
31. Domarus, 1371. The German version of the Soviet request: "Eine entsprechende Verflichtung, ubernimmt die Deutsche Reichsregierung hinsichtlich der in ihren Interesengebieten ansaessigen Personen ukrainischer und weisrussicher Abstammung."
32. See: letter from Chekmeniev in the Appendix "Ruthen" – a term used to define the Ukrainians in eastern Galicia. This was the term used by Austrian government institutions and by the Germans. The Poles referred to them as 'rusini'.
33. See: Hans Schafranek, *Zwischen NKWD and Gestapo. Die Auslieferung Deutscher und Oesterreichischer Antifaschisten aus der Sowietunion an Nazideutzchland 1937-1941.* ("The Handing over of German and Austrian Anti-Fascists from the Soviet Union to Nazi Germany 1937-1941), Frankfurt a/Main 1990, 61.
34. V.A. Nevezin, "Metamofozy Sowietskoi Propagandy v. 1939-1941", *Voprosy Istorii* 8 (Moscow, 1994).
35. Ibid., 166.
36. F. I. Firosov, "*Archivy Kominterna I vneschnaia politika SSSR v 1939-1941; Novaia I noveischaia Historiia, 1992/6,* 18. Hereinafter: Firosov; *The Diary of Georgi Dimitrov 1933-1949* (New Haven, 2002), 115-116.
37. *The Diary of Georgi Dimitrov*, 18 ff.

38. Firosov, 26.
39. Gennadi Kostyrchenko *The Antisemitism of the Soviet Regime from the Beginning to its Peak, 1938-1953* (Moscow, 2005).
40. *Dimitrov and Stalin 1934-1043. Letters from the Soviet Archives.* Alexander Dallin and F.I. Firosov, (eds.), (New Haven, 2000).
41. Ibid., 156.
42. See: Renee Poznanski, *Being a Jew in France* 1939-1945 (Jerusalem, 1999) [Hebrew]. Hereinafter: Poznanski.
43. *Letters from Dimitrov to Stalin,* note 40, 160.
44. Ibid., 174.
45. Poznanski, 83-86. Tuvia Frieling, "*Who are you Leon Berger?, The Story of a Kapo at Auschwitz: History, Politics and Memory*" (Tel Aviv 2009). See Chap. 7, 73-82 [Hebrew].
46. Frieling, 81. The affidavit submitted by Leon Berger (Eliezer Gruenbaum) to the committee of the Polish Workers Party (PPR) convened to discuss his application for re-admission to the Party in Poland. Copy of the document filed in the Ben-Gurion Heritage Archives.
47. Martin Gilbert. *A History of the Jews of Europe during the Second World War* (New York, 1985), 309. Annette Wieviorka on the question whether the Communists were aware of the fate of the Jews, *Deportation et genocide – entre la memoire et l'oubli,* (Paris, 1982), 53-57.
48. This estimate was probably made by the relevant bureau – either of the Soviet Foreign Ministry or of the Chairman of the Council of Commissars which was chaired by Molotov. The latter seems to me more likely as the letter is addressed to him as Chairman of the Council of Commissars.
49. The numbers here relate to the correspondence of the committee led by Chekmanaiev.
50. So long as the German letters do not come to light, we will not know whether the Germans actually offered the Western Ukraine, as it is known today, and not the Southern Ukraine. For in the context of Birobidzhan as a Jewish province, it is reasonable to suppose that this actually referred to Southern Ukraine, since in the twenties the Soviets had plans for a large Jewish population to settle there, and indeed four Jewish districts were defined there. It is also possible that the Germans had in mind the same part of the Ukraine, from Odessa northwards, since in the pre-WWII Soviet Union, this area was known as the Western Ukraine.
51. These apparently are the German appeal papers that were not found among the papers in Molotov's estate, but that might be kept in the archives of the former Russian Foreign Ministry. My attempts, some years ago, to get access to these archives were in vain.

Translated by Miriam Shimoni

Irena Cantorovich

Jews and non-Jews in the Partisan Movement in Belorussia, 1941-1944

The partisan movement in Belorussia was the first such movement in the entire Soviet Union due to the fact that Belorussia was the first area to be occupied by the Germans. The partisan movement was made up of individuals of different nationalities – Belorussians, Russians, Poles, Jews and others, who fought passionately side by side for freedom, for their country and for themselves. In this paper, emphasis is placed on mixed partisan units in which Jews fought alongside non-Jews. Due to the multi-national character of the Soviet partisan movement in general, and of the Belorussian movement in particular, it was only natural that tensions based on ethnic background would arise within these units. But to the same extent, one might have expected that, given the precarious situation facing the entire Soviet Union during World War II, and the fact that everyone was fighting the common Nazi enemy, age-old conflicts and prejudices would be set aside.

The Nazi Occupation of Belorussia and the Behavior of the Local Population

On the eve of the German invasion of the Soviet Union on June 22, 1941, there were over one million Jews living in Belorussia, including refugees from Poland.[1]

In the interim period, from Hitler's rise to power in 1933 to August 23, 1939 – the date the "Molotov-Ribbentrop Pact" (the non-aggression treaty between the Soviet Union and Germany) was signed - reports of the persecution of Jews in Germany occasionally surfaced in the Soviet media. However, after the Pact was signed all such reports suddenly vanished. Consequently, Soviet Jews knew very little, if anything at all, of the fate of their fellow Jews in areas under German occupation.[2] Most of them believed that fascism was the enemy of communism alone, and knew nothing of the effect it was to have on the lives of the Jews. Furthermore, the older generation of Jews, who remembered life under German rule during World War I, maintained that life under the Soviets was even worse, and therefore refused to believe that their lives were under threat in the present situation.[3]

However, it was the secret addendum to the Ribbentrop-Molotov Pact that was to seal the fate of Belorussia and its Jewish population for the next fifty years. On September 1, 1939, Germany overran Poland, its forces advancing eastwards at lightning speed. On September 17, in accordance with the Pact, the Red Army advanced across the Polish border from the East, annexing eastern Ukraine and the western part of Belorussia to the Soviet Union. Right from the start, Germany had no intention whatsoever of honoring the Pact in its entirety, and on June 22, 1941, German forces launched their major offensive on the Soviet Union. By June 25, they had advanced to within twenty km. of Minsk and rapidly overran all of Belorussia.[4] The Germans regarded the war on the USSR as an ideological and racial war, its purpose being extermination and colonial occupation. Once they had completed the invasion of Belorussia, the Germans changed its administrative division, reduced its size and determined that the General District of Belorussia (*Generalbezirk Weißruthenien*) was to comprise the province of Baranovich, the western and north-western districts of the Minsk province, and several sections of the Brest and Vileyka provinces.

The rear support of the Army Group "Center" was set up in Eastern Belorussia, while the administration and large-scale forces were established in Minsk. Until the spring of 1942 the Germans had recruited former Polish citizens as the basis of their administrative hierarchy and garrisons, relying on the public's enmity toward the Soviets. The German regime appointed village and city mayors from among the local population and established police forces based on local volunteers. Assistants were also to be found among the Belorussian migrants in Berlin and Prague.[5]

Under Nazi policy, the Jews were destined for extermination. The platform of the German National Socialist Party defined the annexation of territories in the east as the expansion of 'Lebensraum' for the 'Thousand-Year Reich'. Hitler believed there was a direct link between the Bolshevik regime of the Soviet Union and the Jews, and regarded the Soviet state and Communist ideology as tools wielded by the Jews in order to gain power over the world. Consequently, he perceived the physical elimination of the Jews as the essential path towards the destruction of the Soviet state in preparation for German settlement within its territory. War on the Jews and presentation of the Soviet state as a Jewish-Bolshevik state formed an important part of the German propaganda addressed to the Germans, the Soviet population, the Red Army and the Soviet partisans.[6]

The Jews of Belorussia learned a little of what awaited them from refugees who came streaming in from Poland,[7] who told them of the fate of the Jews under the Nazi regime. These stories and rumors were greeted with disbelief, particularly given the fact that no such reports ever appeared in the Soviet press. In the summer of 1941, for example, rumors were circulated in Malech that German units, together with the

Lithuanian police, had surrounded Pinsk, rounded up all the local Jews and shot them dead. The assignment given the German troops, it was rumored, had been to 'cleanse' the small towns and villages of the area of all Jews. Yet no one believed this story. Even when local police officers leaked rumors of this nature to their Jewish neighbors, they were not taken seriously. However, this apathy was short-lived. Peasants returning from the urban centers brought ever more horrifying stories of pogroms against the Jews in various locations.[8]

The only elements able and permitted to take any initiatives in the Soviet hinterland were the Communist Party and the authorities. As far as they were concerned, it would have been unthinkable to decree the evacuation of a town prior to the launching of a German offensive, lest this be taken as an indication that the Soviet Army was not ready for war. Even when the evacuation of a town was finally organized, it was not carried out all at once. Priority was given to the wounded and to the machinery of the most vital factories, to selected livestock, etc. Jews who happened to work in those factories were evacuated, together with the machinery and with their immediate families, and it is to this that they owe their lives. When at last it was decided to evacuate the rest of the population, the first train to leave – which in many cases turned out to be also the last– was reserved for Party members, civil servants, etc. Jews were included in this evacuation only if they fell into one of the above categories, and not because they were in greater danger. The Soviet authorities did not acknowledge the fact that the Jews were under special threat, even though it appears that from the summer of 1941 they knew of the Nazi policy to eliminate the Jews.[9]

There are many testimonies relating to the large-scale flight of Jews from the towns of Belorussia, right from the start of the occupation. However, these refugees were stopped in their tracks by Soviet officials who refused to allow them to flee to the East; thus they were forced to go back.[10] In other cases people who tried to leave their jobs so as to get out of the danger zone were labeled 'cowards' and were made to stay. And yet the leaders of the Republic were among the first to flee from Minsk, having been evacuated from the city already on the fourth day of the war. In fact the entire population, Jews and non-Jews alike, were left to deal with their bitter fate on their own, with no information available on developments at the Front and no instructions for action.[11]

When the Germans launched the invasion of the Soviet Union, four *Einsatzgruppen* began the massacre of Jews and other 'undesirable' elements. Group B[12] operated in Belorussia. The policy of the occupation regime pertaining to the "Jewish question" in Belorussia was implemented first through the establishment of 164 ghettoes in 153 localities. Some towns had several ghettoes: for example, in Homel there were four and in Zhlobin – six. The Germans viewed the ghettoes as a means of rounding up the Jews so as to facilitate their extermination. In each ghetto, the Jewish community leaders were immediately identified and were the first to be murdered, so as to reduce the likelihood of an uprising by the Jews. The ghettoes in the Soviet Union, as in Belorussia, were set up simultaneously with the start of the massacres, and in some cases even after some of the local Jews had already been murdered.[13]

Some of the ghettoes in Belorussia were liquidated within a short time of their establishment – this began already in the autumn of 1941, i.e. only a few weeks after their formation. In most cases the ghetto would be surrounded by German troops and local police. The Jews were rounded up in the central square and from there were made to walk, or were taken by truck, to the site of the massacre. The majority of the Jews of Eastern Belorussia had been killed by the winter of 1941, and by the latter half of 1942, all the Jews of the western regions had been murdered. From the beginning of 1943 to the withdrawal of the German troops, all the Jews still surviving in the camps and the ghettoes had been killed. The only ghetto still existing in Belorussia by October 1943 was the Minsk ghetto, where the Army Group "Center" had established its base. It is due to the fact that the Jews there were employed in a variety of jobs that they were able to cling on to life a little longer.[14]

The German units responsible for implementing the policy of exterminating the Jews not only faced a chronic manpower shortage, but were also assigned to operate in unfamiliar territory. In many cases they could not identify the Jews among the local population. That is why the cooperation of the population and local officials was of such great importance to them. The district governors, who had been appointed by the Germans from among local residents, participated in the registration of the Jews and their confinement to the ghettoes. On October 22 1941, for example, the Germans in Gluboki announced that all Jews must move to the ghetto within a matter of hours. Local police took charge and beat people with their rifle butts, clubs or whatever came to hand.[15] On November 4 1941, the round-up of all the Jews of Bobruisk in the local ghetto was launched. Jewish property was looted immediately. Just three days later, on November 7 (the anniversary of the October revolution), a massacre was perpetrated. Orders were issued over loudspeakers for all the Jews to congregate immediately by the gate, yet no one showed up. The response of the Belorussian police and German troops was to burst into the ghetto with their dogs. All the captured Jews were loaded onto trucks and driven out to trenches that had been dug in advance, where they were shot. Some days later, the possessions of local Jews could be found offered for sale in local shops.[16]

Upon the withdrawal of the Red Army, local police units in the Soviet area were formed spontaneously. Among their ranks were to be found nationalists, anti-Soviets and anti-Semites, and the German occupation authorities "adopted" these groups as an auxiliary force. Officers of the 11th and 12th police reserve battalions, which were composed of Lithuanian collaborators, displayed the most extraordinary cruelty against Belorussian Jews.[17] During the massacres, local police officers took part in the expulsion of Jews from their homes, in rounding them up and transporting them to the killing ground and guarding the site so that no one could escape. In fact, in many cases it was men from their ranks who actually pulled the trigger.[18]

In the town of Krivichi, local police officers rounded up Jews every day and beat them with clubs. Together with the Germans they searched everywhere for any Jew who might have gone into hiding. In Liozno, on winter nights in 1941-42, Belorussian police broke into homes in the ghetto, smashed windows, beat up Jews and drove them out into the piercing cold. On May 8, 1942, the Lida ghetto was surrounded by German soldiers. The Jews were driven out of their homes by Polish, Ukrainian, Belorussian and Lithuanian police officers. After gaining experience at the Pinsk ghetto, which had been set up on the Passover of 1942 and eliminated within less than seven months, several Polish police personnel participated here too, and were noted for their brutality. On October 14, 1941, 54 Jews were murdered in Kurenets . While it was the Germans who organized the massacre, the actual murders were perpetrated mainly by local people. On the Purim of 1942, Belorussian police murdered another thirteen Jews. And in the ghetto at Lakhva, Ukrainians and Lithuanians took part in driving Jews out of their homes and bringing a group of them to trenches that had been prepared in advance.[19] There were many such manifestations; the above are only a few examples.

Any gentile who concealed a Jew was thereby endangering not only his own life but that of his entire family: if he was caught, he and his family were put to death. Similar regulations applied not only with regard to Jews: even a person who refused to assist Jews but agreed to help escaped POWs or soldiers who found themselves to be under siege would be punished no less severely. However, in the eyes of the local population, any such soldier or ex-POW seeking refuge would be regarded as 'one of us', whereas the Jews - even neighbors and long-term colleagues - were perceived as foreigners. When a local person agreed to shelter a Jew, it would most likely be for payment and once the money ran out, that would spell the end of the hospitality. But no such payment was demanded of non-Jews who were given shelter.[20]

The Germans threatened the local farmers that all their property would be seized if they in any way assisted Jews, and anyone who saved a Jewish life would be executed along with his family. In many cases that indeed was what happened. On the other hand, rewards were promised to anyone who handed over a Jew to the Germans. People of good will who would have liked to help Jews feared to do so in case a neighbor, often acting out of some personal motive of resentment, informed on them. The German occupation authorities also resorted to making false promises, thus raising hopes among the local population. That is why in many locations the occupying force was given a traditional Russian welcome –bread and salt.

The first German units to enter Kurenets on June 28 1941 were welcomed by the local non-Jewish residents with milk and flowers.[21] "The German invasion opened up numerous channels of violence for the anti-Semites, and this was reflected in the views of extensive segments of the population and in their collaboration in the implementation of the Nazi policy towards the Jews."[22]

Prevalent among the non-Jewish population was the view that, "the Jews as a group are to blame for something and therefore must be punished, and some of the Jews undoubtedly deserve to die". That is why many of them assisted the Germans by providing information on the whereabouts of Jews and their property. In some places the Germans boarded up abandoned Jewish property and sold it to locals in exchange for food and everyday essentials.[23]

Most of the non-Jews were indifferent to the fate of the Jews. Only a few individuals expressed misgivings over the suffering being inflicted on them. However, there were instances of cooperation between local non-Jews and the Jewish underground in the ghettoes. Thus, for example, a group of non-Jewish women in Minsk assisted in the smuggling out of some seventy Jewish children from the ghetto, and their transfer to orphanages in the city. We also know of priests who rescued Jews and helped them to reach the partisans. One group among this extraordinary minority was the Baptists, who themselves were a persecuted minority. They believed that the Jews were God's Chosen People and assisted them wherever they could.[24] One of the Baptists who came to the aid of Jews was a Belorussian woman who sheltered a Jew for two years – until a police officer moved into her home. The Jew was discovered and shot dead.[25] A resident of Kossovo concealed 28 Jews in dug-outs which had been used as a tannery for two weeks, and then escorted them to the forest. He was discovered and his entire family was executed. An entire village in the Koidanov area was burned down together with its inhabitants for concealing Jews.[26] There are too many cases of non-Jews in Belorussia coming to the aid of Jews for us to be able to record them all.[27]

Formation of the Partisan Movement in Belorussia

Before we examine the participation of Jews in the Soviet partisan movement, we must first look into the nature of that movement, its organization, its combat system and its tactics.

After the German offensive, the Council of Soviet Commissars (i.e., the government) prepared a work plan for the enemy hinterland which included the formation of partisan and sabotage units in the occupied territories. On July 3, 1941, Stalin gave his first radio broadcast address after the German onslaught. He called on the public to form partisan battalions and sabotage units so as to set fire to bridges, forests and storage facilities, sabotage roads and communications, report on enemy movement and detect collaborators. Orders were also issued regarding the organization of popular resistance. In these decrees, it was clearly stated that, where there was German occupation, the resistance battalions would switch to partisan methods of combat and initiate the sabotage of the means of communication and transportation, supply facilities etc.[28]

This appeal met with widespread response from the local population and partisan battalions, sabotage units and underground cells were formed. In July-August 1941, clandestine units were formed in 98 areas in eastern Belorussia, in both occupied and non-occupied areas; in the latter, they underwent special training, after which they either stayed on in the territory assigned to them before the occupying force arrived, or relocated to the enemy hinterland. By mid-August of 1941, there were no less than 231 such units in Belorussia; however due to the heavy losses suffered by the partisans in the autumn and winter of that year, many were forced to cease their activities. It soon became evident that without a regular supply of arms and munitions, provisions and communications facilities, and in the absence of instructors in various fields of expertise, there would be scant chance of success in waging a partisan war.[29]

Individuals of various nationalities, members of the communist party and its youth movement, civil servants, police officers and many others joined the ranks of the partisan movement in Belorussia. In many cases they did so because they had not managed to retreat with the forces or had remained in the area for some other reason. Red Army officers and men who had not been taken prisoner were among those who joined the partisan ranks. There were groups trained in the Soviet hinterland who transferred to the partisans along with their weapons and equipment, young people who had refused orders to move to Germany to work there, POWs who had escaped to the forests and local people who had served in the German army and police and feared the punishment that awaited them for their collaboration after what was believed to be imminent Soviet victory.[30]

Most of these people had undergone no training for underground or partisan activity. Many were captured and executed by the Germans already in the early weeks of the occupation. From the summer of 1941, partisan activity took place mainly in the forest-covered area of eastern Belorussia, whereas in the western part of the republic, the soldiers and communist activists who had remained behind in the German hinterland were mainly preoccupied with the struggle for mere survival. Until late in 1942, the partisan movement had to cope with a severe shortage of men, equipment and munitions. Some of the battalions did not have enough weapons for all of their men, and even the weapons they did have were for the most part primitive and ineffective in battle against the German troops. Military action was on a limited scale and the main concern was with surviving from day to day and the maintenance of equipment. Only gradually did partisan action begin – first with the burning down of the large farm estates that were the main source of food supplies for the German front, followed by the sabotage of factories that were of economic benefit to the German offensive. Another field of operation was setting ambushes for small groups of German troops and police, etc.[31]

Until May 1942, the Soviet partisan movement had no central command. Most of its units were organized spontaneously and operated independently. What they had in common was the struggle for their very existence.[32] Only on May 30 of that year did the State Committee for Security issue a decree on the establishment of a central headquarters of the partisan movement under the command of Panteleimon Ponomarenko, at the time First Secretary of the Central Committee of the Belorussian Communist Party. At the same time, work was begun on the establishment of regional commands and on September 9, 1942, the Belorussian command headquarters was formed. The central command of the partisan movement formulated and laid down the principles for the tactics and organizational structure of all of the partisan battalions.[33]

Local farms were the most important source of supplies for the partisans, particularly those of families who had collaborated with the occupation regime. These were requisitioned. However, a strict ban was imposed (though was not always adhered to) on the looting of any luxury items or anything else that was not strictly needed for life in the forest, thereby indicating the desire to maintain good relations with the local population, who also happened to be an important source of information. The villagers managed, particularly in the initial period, to gain possession of a significant proportion of the weapons left behind by the retreating Soviet forces, and these turned out to be a vital source of equipment for the partisans. However, the villagers were not always ready to hand over these weapons voluntarily, and in some cases the partisans had no choice but to resort to threats and coercion. Some battalions had special units that specialized in 'finding' and adapting weapons. Workshops were set up for the adaptation and repair of weapons and other means of combat. In some locations, Soviet troops had secreted weapons in caves or stashed them in river banks. And of course arms were seized in the wake of operations against the Germans.[34]

The partisans' principal assignments were to sabotage and destroy the Germans' transportation routes and blow up trains carrying weapons, munitions and troops. Many battalions chose to set up their bases close to traffic routes. Since the Germans were short of manpower for guarding the trains, the partisans attempted – and in many cases succeeded – to derail trains carrying troops, munitions and other equipment. Their numbers reached the thousands: the "railways war" was waged all over the Republic, disrupting German rail traffic often for days on end. This could be called the main achievement of the partisans in Belorussia.[35]

Several regiments were assigned by the central command from the outset to engage in special missions such as espionage, organizing underground units and activists in the cities; sabotaging economic enterprises; assassinating leading German personalities and collaborators. Another important assignment was the capture of Germans as sources of information regarding the plans and deployment of enemy troops.[36] Others were assigned to propaganda, including 'conveying to the local population the voice of war-torn Russia, wounded but still fighting." They seized equipment from German printing houses (often with the help of Jews) and used it to print leaflets and clandestine newspapers for the local population. They disseminated information on developments at the Front and on German policies, in order to dissuade local citizens from collaborating with the Germans. Naturally, these leaflets contained no reports on setbacks suffered by the Soviets in the fighting. The partisans also gave talks to villagers, hoping to persuade them to join their ranks.[37]

The tactics adopted by the partisans was determined both by their immediate targets and by the balance of forces between them and the Nazi military machine at the time.[38] Generally, the partisans would pinpoint the target of their attack, endeavor to carry it out swiftly and withdraw as quickly as possible. The idea was to take the enemy by surprise with sudden, unpredictable raids at unforeseen times, conducted by groups of different sizes. Guards would be posted along all routes leading to a partisan camp and every day a new password would be adopted. Sometimes mines were laid on roads which motorized enemy units were expected to use. The partisan camps changed location frequently, adopting particularly cautious tactics in the relocation. Whenever partisans entered a house, they would post a guard outside. They were particularly wary of local farmers, always trying to ensure that they were kept in ignorance of the partisans' whereabouts. Entrance to partisan camps was permitted only if one could state the current password.[39]

At the beginning of December 1941, the German advance was halted on the Moscow Front and the Soviet counter-attack was launched immediately. The Germans were forced to retreat to the west, to several km. from Vitebsk. "The Vitebsk Corridor", which emerged as a result of the retreat, enabled partisans in the area to seize equipment and supplies. The areas from which the Germans withdrew were from that point on known as "partisan areas". From the autumn of 1941 until November 1942, nine areas came under this appellation in Belorussia – areas that neither the Germans nor their collaborators dared approach for a period of many months.[40]

On June 23, 1944, the Red Army launched the counter-attack that brought about the liberation of Belorussia. Even before this offensive, closer links had been formed between regular units of the Red Army and partisan units, and when the major offensive on Belorussia was launched, it was the partisans who undertook to block roads and highways to the retreating German troops, who found themselves having to fight partisans rather than combat units of the Red Army.[41]

The Path of Jews to the Ranks of Partisan Units

The Jews, knowing that their fate was sealed, tried by every possible means to escape from the ghettoes and the camps, but there were few escape routes available. In any case, it would have been virtually impossible to get away without the tacit cooperation of local residents, without feeling confident that they were not going to be handed over to the Germans or the local police. In all the areas of the Soviet Union under German occupation, there was no organized support for the Jews, neither on the part of the underground nor of any other formal or informal organization, or officials. As far as can be ascertained, neither the Soviet government nor the Communist Party ever gave any instructions to the underground in the occupied areas to offer such assistance.[42]

The conditions for joining a partisan unit generally included fitness for combat, reliability and, in rare instances, possession of a weapon. But obstacles were placed in the way of Jews who sought to join a mixed unit, with particular emphasis on the requirements of fitness for combat and possession of a weapon. These strict requirements were apparently due both to prejudice and to the realization that Jews who fled to the forests did so not as a matter of choice, as was the case with many non-Jews. Indeed, in some cases Jews were admitted to the ranks only after proving themselves by means of test-case operations. These conditions deterred many Jews from applying to join the ranks of the partisans; they also resulted in forcing many of those who had decided to flee from the ghettoes to undertake daring missions to obtain weapons.[43] It is difficult not to regard this as a policy arising from prejudice.

Before the war, the Soviet regime had already dismantled Jewish national institutions.[44] Consequently, when the Germans invaded the USSR, there were no functioning Jewish organizations or Jewish leadership. However, several Jewish organizations that had continued to function clandestinely during the Soviet period formed the nucleus for underground activity after the German invasion.[45]

In western Belorussia alone, underground organizations were formed in some 63 ghettoes and camps, and in some of these riots were planned and even carried out. On the other hand, in the areas to the east, excluding the Minsk district, all able-bodied men of military service age were called up in the early days and weeks of the war. Here, too, the evacuation of factories and institutions was more intensive: they were evacuated together with their workers and their families. The Jews left behind in these areas were mainly women, children, the elderly and the handicapped. There was thus scant possibility that any kind of resistance could be initiated.[46]

Due to the tough conditions prevailing in the ghettoes, it was almost impossible to organize resistance in them, which is what led to the idea of escaping to nearby forests. To many Jewish underground organizations, this seemed the only possible course of action. From the day these clandestine groups were formed to the day their members escaped to the forests, the cells in most of the ghettoes were preoccupied with preparations for escape.[47]

One of the toughest problems was how to procure the weapons which were needed both inside the ghetto for self-defense and outside it, as a condition for joining the partisans in the forests. The principal source of weapons were the arsenals where the Germans hoarded armaments they had seized from the Soviets. Some Jews, including women, were employed by the Germans in arms factories or at German command posts, where they were able to steal weapons to smuggle into the ghetto – at great risk. There was also the possibility of purchasing weapons from peasants who had concealed them after the withdrawal of the Red Army – in some cases this was done through a non-Jewish go-between. However, this meant taking a tremendous risk: on more than one occasion the go-between betrayed the Jews involved, informing on them to the Germans or the local police. And even after weapons were procured, they had to be smuggled into the ghetto – an undertaking that was no less risky. The ghettoes were under constant, stringent security watch, and if weapons were discovered on any premises, the lives of all those living there and indeed – of everybody in the ghetto – would be endangered.[48]

> There were several problems that confronted Jews who decided to flee to the forests:
> 1) the risk of coming face to face with anti-Semitic troops and gangs of various kinds;
> 2) the risk that non-Jewish partisans would seize their weapons and refuse to accept them to their ranks;
> 3) the risk that, even if they were accepted by a partisan unit, they would be assigned impossible missions from which there was no return;
> 4) the risk of being murdered by one of the partisans; and
> 5) the risk of facing execution by the partisans for no reason whatsoever.[49]

These were the dilemmas that faced young, able-bodied men from the ghettoes who could have left, but would thereby have left their families without any support. Such a course of action would have been unacceptable to most of the ghetto population, who had been brought up on traditional Jewish values where the family always came first. In the forests there were no facilities for looking after young children or old people; that is why many gave up on the notion of joining the partisans. While some parents encouraged their children to flee, others were firmly opposed to their trying to escape from the ghetto.[50]

In addition, the German authorities of the ghettos tried their utmost to prevent any revolt within the ghettos, as well as any escape attempts. In the Minsk ghetto, for example, every Sunday the Jews were forced to report at 'Jubilee Square' – one of the central piazzas in the city – which was situated inside the ghetto. They were ordered to reveal whatever they knew about partisans in the area and to inform on anyone in the ghetto who was collaborating with them. The ghetto commander would stand on a table placed in the middle of the square and warn the Jews not to even think about attempting to escape: even if they did make it to the forests, he said, they were bound to die of cold and hunger. But above all, he cautioned them, the partisans detested Jews and were liable to kill them.

Then there was the matter of collective responsibility, a notion which the Germans used to instill fear into the minds of anyone so much as contemplating escape to the forests. In Minsk, the Jews were required not only to sew yellow patches on their clothes, but also a white patch bearing their personal number and address, the idea being to drive home the risk of collective punishment for all the fellow residents of their buildings as a consequence of violating the orders of the authorities.[51]

Escape from the ghetto was only one aspect of the struggle for survival. In many cases, even if a Jew managed to get out of the ghetto, there would be nowhere for him to go. Most of the local residents would be afraid to help such a person due to the risk to themselves and their families. In fact, it was the poorest farmers (old people, landless peasants and widows), whose houses stood isolated, remote from any neighbors, who were most likely to help out by providing food and lodging. Some of those who escaped found refuge with non-Jewish people they had known before the war.[52] The most successful escapes were by organized groups with a guide who was familiar with the area. But in many cases the non-Jewish guide turned out to be unreliable. In some cases the guide was a Jew whose appearance did not give him away. Many of the guides were women, adolescents or even children.[53]

It must be admitted that in many cases Jews from the ghetto were deeply disappointed with their contacts with the partisans, some of whom were dubious characters who exploited the naïve trust placed in them by the Jews and took advantage of them to rob them, seize their weapons or even murder them. This is what happened to ten people who left the Lida ghetto early in 1942 to search for partisans: they encountered a group of armed non-Jews who promptly relieved them of their weapons.[54]

The deep-seated hostility to Jews by the local population was expressed in all manner of ways: robbery, murder and betraying them to the authorities. That is why the Jews tried wherever possible to move only by night,[55] keep to isolated country lanes and avoid the villages themselves. "We knew that we would have to advance very slowly, avoiding all the villages along the way … as the Nazis were not our only enemy and potential murderers. Any gentile whom we came across might be profoundly anti-Semitic. If he was not able to kill us with his own hands, he might lose no time in summoning Nazis to pursue us."[56]

Nevertheless, as we have mentioned, there were individuals who came to the aid of Jews making their escape, thereby taking the risk of being themselves sent to a concentration camp or even of being executed together with their entire family. There were peasants who provided food, and some even gave shelter to Jews for a short period of time.[57] Thus, for example, early in the autumn of 1943, the Mazeritzki couple and their daughters hid in the home of a farmer until his neighbors informed on them and they were shot to death. Only the youngest daughter, Genia Mazeritzki, survived and fled to the forest; she reached the partisans and there was assaulted by one of the men. When she put up fierce resistance, the rapist shot her dead. It should be noted that he in turn was executed for this act by the partisans.[58]

Jews who managed to reach the forests and join the partisans were organized in separate battalions. Camps were set up for the elderly and families from the ghettoes that came under the care of Jewish partisans who would provide food and protection. There were also cases of small Jewish units being integrated into large partisan units in which the Jews were a minority.[59]

Jews who sought to join the partisans were not always willingly accepted. Even people who were fit for battle and for survival in harsh conditions were at first rejected and had to negotiate at length, resulting in cases of people concealing their Jewish identity due to the prevailing attitude of prejudice.[60] However, despite all these difficulties, the number of Jews who fled the ghettoes climbed steadily – a process that continued until the liberation in the summer of 1944.

The Jewish Underground in the Minsk Ghetto as a Test Case

The underground in the Minsk Ghetto was, as aforesaid, the first to be organized. The first underground units had been organized in the ghetto by August 1941, and in October they merged to form a single unit. Most of their leaders came from western Belorussia. Some of the founders, such as Hersh Smolar, were members of the Polish communist party, and others, among them Nahum Feldman, were members of the local party; and then there were others with no party affiliation. Many of them had come to Minsk as refugees and had a wealth of political experience behind them, acquired in Poland during the 1930s. The underground in the Minsk Ghetto was organized in groups of ten reliable individuals who were very carefully picked. The Judenrat was among the elements that assisted the underground – its members in Minsk cooperated with the underground and some paid for it with their lives. The declared aim of the Jewish underground in Minsk was to enable as many Jews as possible to leave the ghetto for the forests, and join the partisans. As the Germans continued with the systematic annihilation of the ghetto population, it became ever more urgent to bring out as many able-bodied groups as possible.[61]

Another field of activity of the underground was issuing and distributing information bulletins to the ghetto population and establishing contacts with other underground organizations in the city, and non-Jewish partisans.

With the growing intensity of the mass killings, it became increasingly urgent to escape from the ghetto. "We saw the ghetto manifestly in its death throes and here we were within the ghetto, as people condemned to death. The only way we could possibly survive this nightmare was to reach the forests." It was decided, rather than attempt to start an uprising in the ghetto – an attempt *a priori* doomed to failure – to concentrate on "bringing more Jews out to the forests in as secret an operation as possible, and from there, together with our Russian comrades, to wage armed conflict."[62] The underground in the Minsk Ghetto would send out special emissaries, generally women of non-Jewish appearance, to check out the terrain, seek the most suitable location in the woods and ascertain which of the villagers could be relied upon to help.[63]

The Jewish underground provided assistance for the Belorussian urban underground, particularly in the forging of documents and obtaining clothing and papers. Occasionally, leaders of the Belorussian underground who suspected the Gestapo was on their trail sought refuge in the ghetto. The Jewish units also helped with medical care to the sick and wounded among Belorussian underground combatants, providing urgent medical treatment and surgery when required. The wounded would be smuggled into the ghetto to be treated by doctors who were members of the underground.[64]

Some underground members in the Minsk Ghetto were employed by the Nazis in their weapons stores and managed to smuggle out arms.

Others, who had contacts with civilians outside the ghetto, tried to use those contacts to obtain weapons. One member, Aharon Pieterson, for example, obtained six rifles from a non-Jewish friend he had known from the days of the civil war, when they had fought side by side. Some of the weapons that were obtained proved to be in poor condition and had to be repaired inside the ghetto. The arms would be stored in secret caches - some were concealed in the ghetto's Jewish cemetery. One top secret operation in the ghetto was the manufacture of various types of weapons, such as hand-grenades, knives, etc.[65]

From the first days of the occupation, the Communists began to organize in the city, and one of the first assignments of the ghetto underground was to establish contact with them. In November, 1941 these groups decided to merge. This united group was led by Isay (Yehuda) Kazinets, a Jew from Ukraine who managed to conceal his Jewish identity. Its purpose was to set up partisan bases in the forests and to carry out acts of sabotage and conduct propaganda operations in the city. The initial contact was established by Michael Gebelev, who had been granted a permit by the ghetto authorities to move around outside its perimeters in the context of his work. While the two groups agreed to cooperate, the Communist group was at best reluctant in its response to appeals from the ghetto underground to assist in smuggling Jews out of the ghetto.[66]

At that time, a group of Red Army officers in the city organized to form what was known as the "Military Council". They established contacts with both the urban and the ghetto underground, but they disagreed with the ghetto underground over the question of transferring Jews to the forests. "The Council" preferred its own people, their trained officers and men. In early March 1942, the leaders of the "Council" were arrested and, after being subjected to interrogation, more than

400 members of the urban guerrilla groups were arrested, of whom over 200 were executed. That is also how the Germans obtained detailed information concerning the underground in the ghetto. The heads of the Judenrat who helped the ghetto guerillas were arrested in March, 1942 and executed by hanging. A little while later other activists of the underground were also arrested and killed. Nevertheless, activity in the ghetto continued and in April two groups left for the forests. Minsk Jews who reached the forests from the ghetto formed seven Jewish partisan battalions, six of which gradually became mixed battalions when they were joined by non-Jews.[67]

The Jews in the Forests and the Attitude to Them

Most of the Jewish partisans thus operated in mixed units, many of them outstanding in their combat performance. For example, in the 'Voroshilov' battalion, fourteen Jews were awarded medals while still in the forests.[68] In the Soviet partisan movement, many Jews held command positions. The commissar of the battalion known as 'For Soviet Belorussia' was Jewish, as was the Commissar of the 'Stalin Brigade' in the Naliboki forest. Jews also proved outstanding as commanders of troops and sabotage units, and were awarded medals in recognition of their service. Many partisan brigades and battalions were multi-national in character. For various reasons, but mainly because of the Antisemitism rampant in the forests, many of the Jews in command positions preferred to conceal their origins.[69]

In most cases, Jewish commanders were in charge of Jewish units or units with a substantial Jewish majority, but there were some cases of Jews being in command of units in which the Jews constituted the minority. Two Jews in Belorussia were in command of partisan brigades, four Jews acted as brigade commissars, seven Jews were at the head of brigade central command and eighteen commanded partisan battalions.[70]

Mordechai Gurevich, a member of the Baranovich ghetto underground, commanded a group of saboteurs and was attributed with the destruction of twelve trains. He was subsequently promoted to commander of the company, which was composed almost entirely of non-Jews. David Malovitski, also from the Baranovich Ghetto, was outstanding in sabotage operations and was appointed commander of a platoon composed of both Jews and non-Jews which carried out a large number of operations.[71] The Parkhomenko unit numbered 130 members, of whom 85 were Jewish. Its commander-in-chief, its platoon and company commanders and its Politruk were all Jewish.[72] In the Zhdanov battalion, the Politruk was also Jewish, and forty German officers and men were taken prisoner under his command.[73] In the Kirov Battalion (No. 537), the deputy, Politruk, and secretary of the party cell were both Jewish.[74] Vladimir Shatsman was the Jewish commander of a battalion in the Kobrin district. He participated in 17 major battles and many clashes with the Germans and their collaborators. David Rubin was commander of a detachment of 15 saboteurs from the 'War for the Homeland' battalion, which was posted in the Minsk area.[75] There were many others like them.

The partisan units in the forests also had to cope, throughout the entire period of their activities, with disease and injury among their ranks. They relied on the Jewish medical workers (doctors, nurses and paramedics) who came to the forests for treatment. There were occasions when the partisans sent special emissaries to the ghettoes to bring out any surviving physicians. One such instance occurred when a group of Jewish partisans from the Lipichany forest was assigned by its commander to enter the Lida Ghetto to fetch a surgeon, Dr. Miasnik, the idea being that, together with a group of other Jewish doctors, he would set up a field hospital. Here medical care was also provided for partisans from other districts. Jewish girls quickly mastered the arts of nursing and hygiene and helped with the care of the sick and wounded.[76]

The Jewish doctors also offered medical care to the village population, and many participated in the fighting. Almost all the physicians in the battalions of the Kirov Brigade were Jews. Most of them had been called to service by the partisans and were smuggled out of the ghettoes together with their families. In the multi-national 'Iskra' battalion, three doctors and a dentist, as well as two of the three paramedics, were Jewish. A physician in the Voroshilov Brigade was Jewish. He seems to have had considerable influence on the battalion commander and occasionally was able to curb manifestations of Antisemitism.[77]

In many of the partisan battalions, Jews practiced their civilian trades, such as shoemakers, carpenters, cooks, bakers, fitters, mechanics and electricians. Some of them played an important part in the assembly of weapons, laying of mines etc. Jewish tanners were able to help solve the footwear problem, fashioning soles out of the hides of animals slaughtered as food for the partisans.[78]

A good number of Jews were also employed in the interrogation of German captives, in the decoding of documents found in the wake of attacks on German expeditionary forces and among the prisoners' belongings. They also translated articles from the German press. In many instances Jewish doctors acted as interpreters. Another important role filled by Jews was the organization and distribution of partisan publications. They were also able to provide important equipment needed for the partisan printing presses – equipment they had stashed away while still in the ghetto and employed in German printing works, and had brought with them when they fled to the forest. This is what happened in the ghettoes of Lida and Minsk.[79]

Many of the Jews who fled to the forests had 'personal accounts to settle' not only with the Germans but also with the local population, with people who had collaborated - whether directly or indirectly - in the murder of Jews. Once they were in the forests, the Jewish partisans did their utmost to seek vengeance on collaborators among the police and on farmers who had collaborated with the Germans and profited from what their victims had left behind. They carried out acts of vengeance that did not always have the support of the non-Jewish partisans, who were anxious to maintain good relations with the local population. Despite this, the Jewish partisans did their utmost to prove to those who had hated and persecuted them that they were still around, and were determined to avenge the blood of their tortured brethren.[80] Many survivors testify about this: "We had little hope of surviving the war but we thought that, so long as we had the strength to do so, we had to avenge the blood of those dearest to us";[81] and: "I did not fear death. What mattered to me at the time was to avenge the deaths of my mother and my sisters: what had they ever done to deserve to be brutally murdered?"[82]

Despite all of the above, the integration of Jews in the Soviet partisan movement was no bed of roses. The partisan HQ, as well as the Soviet authorities, did not react appropriately to the situation of the Jews under the German occupation,[83] partly because they thought that they would thereby be bolstering Nazi propaganda, which repeatedly drummed home the allegation that the Soviet regime was dominated by Jews. Aid to those forcibly held in the ghetto was not mentioned, even in the decree defining "the tasks of the partisan movement" issued by the partisan movements' Central HQ on September 5, 1942.[84]

Admittedly, in January 1943, the Partisan Central HQ drafted a four-page document concerning the Germans' attitude to the Jews, but this document brought about no change in the partisans' approach. In bulletins issued by this HQ, while the civilian population is called upon to come to the aid of families whose sons have been sent to work in Germany or have been murdered by the Germans, there is no mention of coming to the aid of Jews. The exodus of Jews to the forests in order to participate in the partisan struggle or to find shelter and safety was initiated by the Jews themselves, and the Soviet partisan movement did not play any direct or active role in this, or take any initiative towards rescuing the Jews.[85] Many partisans were reluctant to undertake the burden of looking after women, children, the elderly and the sick – people who had the greatest difficulty in adjusting to life in the forest. "Our battalion is a combat unit, not an old-age home," many refugees from the ghettoes were told when they reached the forest.[86] On more than one occasion Jews were suspected of having been sent by the Germans to spy on the partisans and inform on their location, eliminate their leaders and poison wells.[87]

In areas where groups of Ukrainian nationalists or Polish partisans[88] were located, the Jews had virtually no chance of survival. This was the situation in western Belorussia, where Polish partisans were deployed. Many Jews who sought refuge in the forests were shot dead. A similar policy was pursued by some Soviet partisan battalions. This is true for Pavlov, commander of a partisan brigade in the Mohilev district, and he was never charged. Others would rob people they suspected of being Jewish of any valuables they might possess, beat them up and threaten to kill them. Many of those who were not killed were forced to return to the ghetto after the weapons they had acquired with great effort had been stolen from them, together with anything else of value they had managed to smuggle out of the ghetto.[89]

There were also cases of Jews being murdered either en route to an assignment or on returning from it. From testimonies given by many survivors after the war, we learn that they feared fellow-partisans no less than they feared the Germans.[90] Even expertise in a field of importance to the partisan movement was no guarantee for safety. Such was the case of Dr. Avraham Abramovski, formerly head of surgery at the hospital in the Baronovich ghetto, who escaped to the forest in December 1942. On his way he met with a group of partisans, and after he revealed to them that he was a Jewish surgeon from the ghetto and that the case he was carrying contained medical instruments and medication, they stole his case and shot him dead outside the village.[91]

There were frequent cases of Jews being killed by partisans even in 'ordinary' times, but they increased during the German sieges. Their lives became a torment. This did not change significantly after the siege was lifted. "The Jews regard themselves as 'Partisanka' in the forest – i.e. as Jews. They were conscious of their Jewish identity at every step and with every move, and the others made sure they never forgot it."[92] In some battalions anti-Semitic tendencies prevailed to the point that the Jews were forced to flee to another battalion or another district. As the partisan movement expanded, and as Soviet victory appeared more likely, the partisan ranks were joined by increasing numbers of people who had formerly collaborated with the Germans and who hoped this would save their lives. The atmosphere in the battalions which they joined grew increasingly ugly and the situation of the Jewish partisans deteriorated significantly.[93]

In most cases the non-Jewish commanders defended the anti-Semitic partisans in their units and in many cases they were let off after investigation. According to the orders for re-organization, which were issued by the Partisan Central HQ, "it was compulsory for every person or group of people in the forest to belong to a partisan unit." This regulation was interpreted by some of the non-Jewish partisans as according legitimacy to the persecution of Jews and rejection of applications from Jews to join their units. There were Jewish partisans who displayed courage and protested against the anti-Semitic actions of their partisan colleagues. They felt that they were thereby defending their own honor as Jews and the honor of their fellow Jews who had been attacked. But some Jews felt that they had been abandoned by the Command and exposed to revenge.[94]

In the autumn of 1943 the situation of Jews fighting in partisan units took a turn for the worse: the Germans began to bomb villages which they suspected of aiding the partisans and to distribute leaflets with the following message: "The war you are fighting is quite illogical. Surrender. We'll fix you up with good jobs. We will only destroy the Jews, who are to blame for the outbreak of the war."[95] These leaflets served to exacerbate the Antisemitism latent among the partisans.

Even at this stage the non-Jewish partisans continued to place obstacles in the way of Jews seeking to join their ranks, giving the same reasons as before. Women, minors and those regarded as too old to fight were rejected and had to traverse huge distances looking for a brigade that would accept them.[96]

Opposition in principle to the notion of exclusively Jewish partisan units was another important issue. The partisan movement in the Soviet Union was organized on a regional basis or according to the separate Republics. Its commanders were opposed to the notion of independent Jewish battalions, in accordance with general Soviet policy with respect to nationalities, and with the Soviet Union's utter rejection of the German propaganda assertion that the partisan movement was a Jewish enterprise, dominated and run by Jews. This approach not only reflected the anti-Semitic tendencies prevalent among some of the non-Jewish partisans, but also led to the argument that the formation of Jewish brigades entailed the risk of continual conflict with the local villagers, among whom Antisemitism was so deeply rooted. Consequently, the independent Jewish groups were gradually broken up. Men deemed by the non-Jewish partisans to be fit for combat were assigned to mixed units. Reports of the Germans' extermination of the Jews, of the looting of Jewish homes and of the active collaboration of the local population with such actions reinforced anti-Semitic emotions and left the Jews exposed and unprotected.[97]

There were three categories of accusations made against Jews in partisan units: accusations stemming from false reports from the command (i.e. that the Jews were spies); prejudice (that the Jews were wealthy, cowardly and unfit for combat); and using the Jews as scapegoats (where relations with the local population were less than satisfactory, the blame was placed on the Jews).

As early as mid-September 1942, the central HQ of the partisan movement issued a document to partisan units warning them against Jews who claimed they were victims of persecution and were attempting to join their ranks. Such people, according to this document, were actually German agents. In September 1943, the central HQ issued an Order of the Day warning battalions not to accept any new Jewish women into their ranks on the grounds that the German espionage command in Minsk had sent out 200 Jewish girls infected with venereal disease to infiltrate partisan units with the specific purpose of spreading infection. In the spring of 1943 a parade was held in one of the battalions in the Mstislavl district at which the commanding officer read out an order from Central HQ stating that the Germans were sending Jews from the ghettoes to poison the wells and the food of the partisans. In light of these warnings, the partisan leaders were inclined to accept only individuals they knew personally into their ranks. Many of the partisans actually believed that to accept Jews into their ranks meant taking a considerable risk. They thought that a crafty spy would most likely adopt (false) Jewish identity so as not to provoke suspicion among the partisans.[98]

One of the prejudices prevalent among the partisans was the notion that all Jews were wealthy. It was not unusual for Jewish partisans to hear remarks such as: they handed over all their "money, furs and valuables" to the Germans and now they have come to save their skins. On more than one occasion a rumor was circulated to the effect that a Jewish partisan was in possession of valuables which he was concealing. In other cases, when the general order was issued to hand over all valuables, it was pretty obvious that it was addressed solely to the Jews. A Jewish partisan in the Kirov brigade was executed after "he was accused of not handing over all the gold and cloth in his possession." There were times when Jews were ordered to hand over their money, their watches and personal belongings, ostensibly for the purchase of weapons for the battle, or "for the homeland". Often such items were spotted the next day in the possession of those who had confiscated them or of their mistresses.[99]

The myth of "Jewish cowardice" penetrated deep into the partisan movement right from the outset. The Jews, for their part, naturally tried to dispel this notion: "It was my intention to shoot to kill. I will not give my comrades in the battalion any satisfaction of being able to call me a coward." The Jewish combatants in the 'Avenger' brigade displayed courage, skill and a volunteering spirit, but this did not prevent manifestations of Antisemitism against them.[100] Over and over again, Jews were to hear the accusation that they were being slaughtered in their masses in the ghettoes due to their cowardice and their failure to put up any resistance, and that it was due to their cowardice that not all the Jews had fled the ghettoes to the forests to try to form their own mass partisan movement.[101]

The partisans also harbored preconceived notions regarding the Jews' fighting capabilities - notions for which there were some grounds, since many of those who fled to the forests were simply not fit for battle. It was because they had no alternative that they took to the forests, given the danger of total extermination faced by the Jewish community. Some fled with their families, bringing along women, children and old people.[102] In many cases it was said that "they came here with no weapons ... and we will have to protect them, jeopardizing our own lives. First they handed over to the Germans their money, their gold and their diamonds, and then they came here to eat our bread.. They are a superfluous element."[103]

Another accusation made against the Jewish partisans was that they ruined relations with the local population, while the partisan movement considered good relations with them to be of the greatest importance. Commanders tended to set a quota on the number of Jews in their battalions in order not to give the local population the impression that they were specifically defending the Jews – Israel Goldman, for example, who came from Pinsk, was executed by his commander on the charge of looting peasants' homes. The death sentence was also imposed on another Jew who, before formation of the battalion, had taken clothing and food from farmers. It was only due to vociferous protests by fellow Jews in the battalion that his life was spared, but he was then assigned to carry out three dangerous missions.[104]

"There is no one here to complain to, no one to approach."[105] That was the feeling prevalent among many of the Jews who reached partisan units, only to encounter humiliation and discrimination. Many chose to hide their Jewish origin, changing their names and registration details. Faye Schulman recalls that in her brigade there were only a few Jews who did not conceal their Jewish identity. Many of them were outstanding in combat, served as commanders and were killed without their Jewish identity ever being discovered.[106] Rakhmiel Idelson, for example, was 17 when his entire family was confined to the Minsk Ghetto. Fellow members of the urban underground provided him with false identity papers, which is how he was admitted into the ranks of a partisan brigade. It was only thirty years after the war that he received the medals awarded him.[107]

There were, of course, stories of a different nature as well, in which partisans protected Jews and even freed Jews from the ghettoes. In June 1942, the Shchors Battalion arranged for the escape of 170 Jews from the Slonim Ghetto. The same battalion also overcame the Kossovo garrison and enabled 200 Jews to flee from the ghetto. A partisan battalion composed mainly of peasants operated in the Vibetsk region. Back in September 1941, this battalion had established contact with Jews who were working in small towns and villages in the area to form joint sabotage units. Able-bodied young Jewish men, as well as old men, women and children began to arrive at the unit from nearby towns. A small civilian camp was set up for them.[108] Many of the partisans had no interest whatsoever in the 'national question' and received Jews into their ranks without discrimination, relating to them above all as comrades-in-arms.

There were a number of instances of partisan units transferring hundreds of Jews to beyond the front lines.[109] At the beginning of 1943, a six-man partisan unit managed to get 218 Jews out of the ghettoes in the Minsk and Villeika districts, deep within enemy territory – to the "Great Land.[110] In his report to Ponomarenko, the commander of the Kisiliov group wrote that, after the massacre of Jews in those districts and in the city of Minsk itself, those Jews who were able to escape fled to the forests and roamed the area of partisan activity. The Command decided they should be admitted into the existing brigades. In June 1942, after the Central HQ gave its consent, the commanders of the Kisilov battalions were ordered to gather all the Jews and get them out to the "Great Land". They were organized for a march of hundreds of kilometers through areas dominated by German troops and their local collaborators. Nevertheless, all the Jews who set out on this journey in January 1943 were successfully transferred to the Red Army.[111]

Conclusion

Against all the odds, under conditions of German incitement to kill all the Jews, and with a hostile attitude prevailing among the local population, the Jews nevertheless managed to integrate in the partisan movement and play a significant role in the struggle against the enemy. There was no field of partisan activity in which Jews did not participate: be it in combat itself or in medical and other services for the fighting forces. Their professionalism in a wide range of skills was well known before the war and in the course of it, they had ample opportunity to prove it.

Jews from the ghettoes came to the forests after long months of suffering. Many had lost their entire families and now this route appeared to them as their last resort. In many cases, even there, instead of the freedom and calm they had hoped to find, what they got was a hostile reception. They tried to help their fellow Jews by any means possible. In these endeavors they sometimes encountered hostility from their fellow partisans, who did not refrain from voicing their displeasure at the idea of assisting Jews. They regarded those who had escaped from the ghettoes and camps as a burden that further endangered their already extremely risky undertakings. Disputes on this issue became one of the principal motives for the formation of the Jewish camps – the so-called 'family camps', which were open to any Jew who wished to join, even if he came empty-handed and unable to participate in combat.

However in other cases, non-Jewish partisans risked their lives to rescue Jews and take them into their units, or helped in the formation of Jewish units under their protection. Here Jewish partisans showed their resourcefulness, initiative and readiness to sacrifice their lives for their fellow combatants, Jews and non-Jews alike. The desire for vengeance against the Nazis that permeated Jewish thought at the time prompted the Jewish partisans to give their all in the conflict waged by the partisan movement which, in Belorussia, was a multi-national movement. Alongside the Belorussians (who made up approx. 71.1 per cent of the partisans), over seventy other nationalities and ethnic groups were represented in the partisan ranks. In the early years after the war, the Soviet press referred to 200-250,000 partisans. Later on, their number rose to 330-369,000, and research published in the late 60s put the figure at 460,000 partisans.[112]

It is impossible to accurately assess the number of Jewish partisans in Belorussia. As aforesaid, many Jewish fighters hid their true identity and retained their adopted, fictitious identity after the war. Furthermore, before we discuss the matter of numbers, we must seek a precise definition of who is a partisan. For example, could all those who stayed in the so-called 'family camps' but did not take part in the fighting be considered partisans? In addition, it should be borne in mind that registration of the roll-call of a battalion under the chaotic conditions of the forest was not always accurate, and the Soviet policy of deliberately concealing the Jewish origin of partisan fighters put further obstacles in the way of determining the exact number of Jews - a statistic which we are unlikely ever to ascertain.

The Belorussian partisan movement was in fact a miniature model of Belorussian society in general at the time. Among the Belorussian public there were two distinct attitudes to the Jews: the first was the positive approach - an attitude of solidarity, of sharing a common fate and a willingness to come to the aid of Jews despite the risk to one's own life. As against this, there was the attitude of suspicion and hostility that had been prevalent in Belorussian society from long before World War II and that only intensified during the course of the war.

The partisan movement did not think it was under any obligation to add the rescue of Jews to its list of duties, even though its leadership was fully aware of the destiny mapped out for the Jews. As described above, Jews were accepted into the partisan ranks only insofar as they could contribute to the movement with their skills, but not specifically so as to rescue them. The Jewish partisans faced the same challenges as the entire partisan movement, but also challenges specific to them, stemming from their Jewish identity, ranging from the hardships of life in the forests to the daily risk to their lives that haunted them – not only from the Nazis but also from their collaborators among the local population, and even in some cases from their fellow partisans. The Jewish partisan thus found himself fighting on multiple fronts: "the struggle for the right to be a combatant and an avenger, the right to proudly assert one's humanity and the right to a Jewish existence."[113]

> Even in the partisan forests, in the company of fellow combatants who were not Jewish, the Jewish partisan had no easy path. There too he faced numerous obstacles; there too Antisemitism lurked; there too people from the surrounding areas who came to the forests had come under the influence of venomous Nazi propaganda. Even in the forests, Jewish partisans were shot for various reasons and Jews who had sought refuge in the forest were robbed or even murdered. Even in partisan society, the Jew was first and foremost a Jew.[114]

Notes

1. Encyclopedia of the Holocaust, edited by Israel Gutman (Jerusalem, 1990), Vol 1, 173, 176. [in Hebrew].
2. On what the Jews knew or did not know see: Mordechai Altshuler, "Metsukatam shel Yehudey Brit-HaMoatsot BeIkvot Heskem Molotov-Ribbentrop", Yad Vashem 36 (2), 2009, 63-96; Mordechai Altshuler, "HaShoa BeTikshoret HaHamonim BiVrit-HaMoatsot BiYemey HaMilkhama UbaShanim HaRishonot Akhareha – Mabat Khadash", Yad VaShem 39(2), 2011, 107-146. [both in Hebrew].

3. *The Jewish Partisans,* edited by M. Gefen et al (Merhavia, 1958), Vol 1, 503; Yitzhak Arad, History of the Holocaust: Soviet Union and Annexed Territories (Jerusalem, 2004), Vol 2, 966, 968. [Both in Hebrew].
4. Shalom Cholawski, *The Jews in Belorussia* (White Russia) During World War II (Tel Aviv, 1982), 56-60 [in Hebrew]; Alexander Dallin, German Rule in Russia, 1941-1945 (Colorado,1981), 27, 29.
5. Minsk, *Jewish Mother-City* – Memorial Anthology, edited by Shlomo Even-Shoshan (Jerusalem, 1975), Vol 2, 275; Yitzhak Arad, "Shoat Yehudey Brit-HaMoatsot", Yad Vashem 21, 1991, 10; Leonid Smilovitsky, "Yehudim UPolanim Ba-Tnua HaPartizanit HaBelorusit, 1941-1944", *Yalkut Moreshet* 63, 1997, 83-84. [All in Hebrew]
6. Yehoshua Buchler, "Retsakh HaYehudim BeMasve shel Lekhima Neged HaPartizanim BaShtakhim HaKvushim shel Brit-HaMoatsot, 1941-1942", Dapim LeKheker Tekufat HaShoa, Vol 13 (1996); Ben-Cion Pinchuk, "Emtsaey HaTikshoret BiVrit HaMoatsot URedifat HaYehudim BiShtakhim Kvushim Al-Yedey HaNatsim, 1939-1941", Yad Vashem 11, 1976, 172-178; Arad, Soviet Union and Annexed Territories, Vol 1, 169, 174. [All in Hebrew]; Emmanuil Yoffe, Stranitsy istorii evreev Belarusi (Minsk, 1996), 108; Leonid Smilovitsky, Yevrei Belarusi: iz nashei obshchei istorii 1905-1953 (Minsk, 1999),12. [Both in Russian].
7. On the refugees from Poland see: Yosef Litvak, Polish-Jewish Refugees in the USSR, 1939-1946 (Jerusalem, 1988), [In Hebrew].
8. Hersh Smolar, Soviet Jews Behind the Ghetto Barrier (Tel Aviv, 1984), 35-36; The Jewish Partisans, Vol 1, 601; Israel Berestizki, Run for Your Life, My Child! The Life Story of a Partisan and Soldier (Tel Aviv, 2004); Leonid Smilovitsky, "Al HaAyara Glusk", an article that was never published; Yad Vashem Archive, O.3, file 2088. [All in Hebrew].
9. Arad, "Shoat Yehudey Brit-HaMoatsot", 7 [In Hebrew]; Daniel Romanovsky, "Soviet Jews under Nazi Occupation in Northeastern Belarus and Western Russia", Bitter Legacy, edited by Zvi Gitelman (Indiana, 1999), 233-234; Kenneth Slepyan, "The Soviet Partisan Movement and the Holocaust", Holocaust and Genocide Studies, Vol. 14, No. 1 (Spring 2000), Nepokorennaia Belorussia, edited by Vladimir Levdanski et al (Moscow, 1963), 21, 56, 97, 119. [In Russian].
10. Shalom Cholawski, "Minsk BeMaavaka VeKhilyona – KeFetakh LeHavanat Matsavam shel Yehudey Brit-HaMoatsot BaShoa ULeakharea", Yalkut Moreshet 18 (1974), 107-108; Bobruisk Memorial Book, edited by Yehuda Slutski (Tel Aviv, 1967), 734; Yad Vashem Archive, O.3, file 3483. [All in Hebrew]; Villa Orbach, "The Destruction of the Jews in the Nazi-Occupied Territories of the USSR", Soviet Jewish Affairs, Vol. 6, No. 2, 1976, 15-16.
11. Yoffe, Stranitsy istorii evreev, 109, 119-120; Svetlana Rudenkova, "O nekotorykh problemakh organizatsii partizanskikh formirovanii i ikh materialno-tekhnicheskom obespechenii", Staronki vaennaii historii Belarusi, Vol 1, (Minsk, 1992), 98; "Iz vospominanii pervogo sekretaria TsK KP(b)B P.K. Ponomarenko o nachale Velikoii Otechestvenoi Voiny", Belarus, 22 iiulia 1941 goda: govoriat arkhivy, edited by

Rostislav Platonov and Vladimir Fedosov (Minsk, 2001), 131; Galburt Alexandr, Vospominaniia (Minsk, 2000), 11. [All in Russian].

12. Encyclopedia of the Holocaust, Vol 1, 80-88 [In Hebrew]; Orbach: "The Destruction of the Jews", 16, 18.
13. Arad, Soviet Union and Annexed Territories, Vol 1, 255 [In Hebrew]; Leonid Smilovitsky, "Righteous Gentiles, the Partisans and Jewish Survival in Belorussia 1941-1944", Holocaust and Genocide Studies, VII, No. 3(Winter 1997), 303-304; Gennadii Vinnitsa, "Orshanskoe getto", Evreiskii kamerton, 18.10.2001, 9 [In Russian].
14. Yitzhak Arad, "Alfred Rosenberg VeHaPitaron HaSofi BeShitkhey Brit HaMoatsot HaKvushim", Yad Vashem 13, 208; Arad, "Shoat Yehudey Brit-HaMoatsot", 12 [Both in Hebrew]; Romanovsky, "Soviet Jews under Nazi Occupation", 244; Orbach, "The Destruction of the Jews", 22-23; Smilovitsky, "Righteous Gentiles, the Partisans and Jewish Survival", 303-305; Yoffe, Stranitsy istorii evreev, 116-118 [In Russian]; Smilovitsky, Evrei Belarusi, 132, 134 [In Russian]; Emmanuil Yoffe, "K voprosu o kollaboratsionizme v Belarusi v 1941-1944", Evrei Belarusi istoria i kultura, Vol III-IV (Minsk, 1998), 166 [In Russian].
15. Leonid Rein, "Otnoshenie belorusskogo naseleniia k evreiam vo vremia Katastrofy, 1941-1944", Evrei Belarusi istoriia i kultura, Vol VI (Minsk, 2001), 184-187; Black Book, edited by Vasilii Grossman et al (Vilnius, 1993),149. [Both in Russian].
16. Shalom Cholawski, In the Eye of the Hurricane – The Jews in Eastern Belorussia during World War II (Tel Aviv, 1988), 194-195 [In Hebrew]; Victor Liventsev, Partizanskii krai (Leningrad, 1951), 63 [In Russian].
17. Liventsev, Partizanskii krai, 34; Yoffe, "K voprosu o kollaboratsionizme",176. [Both in Russian]; Yehoshua Büchler, "Local Police Force Participation in the Extermination of Jews in Occupied Soviet Territory 1941-1942", Shvut, 4 (20), 1996, 82-82. On the participation of the Lithuanians see for example: Martin Dean, "Lithuanian Participation in the Mass Murder of Jews in Belarus and Ukraine, 1941-44", Rodopi (Amsterdam–NY, 2004), 286-290.
18. Arad, "Shoat Yehudey Brit-HaMoatsot", 13, 25-26 [In Hebrew]; Yad Vashem Archive, O.3, file 1362; Yad Vashem Archive, O.17, file 43; Rein, "Otnoshenie belorusskogo naseleniia k evreiam", 187 [In Russian].
19. Smolar, Soviet Jews Behind the Ghetto Barrier, 61, 63; Cholawski, In the Eye of the Hurricane, 202; Ida and Yosef Kaplan, Ad Yom Kippurim Ze (Tel Aviv, 1975), 46; Fanny Solomian-Lutz, Neara Mul Gardom (Ramat Gan, 1971), 57, 63, 65; Yehuda Bauer, "Kurenets – Ayara Yehudit BaShoa", Yalkut Moreshet 73 (2002), 83 [All in Hebrew]; Black Book, 154-155; V. Shapiro, "Kopel Kolpanitskii iz vosstavshei Lakhvy", Okna 7.3.2002, 22 [Both in Russian].
20. Romanovsky, "Soviet Jews under Nazi Occupation", 239-240; Daniel Romanovsky, "Otnosheniia mezhdu evreiami i neevreiami severo-vostochnoi Belorussii i zapadnoi Rossii", Vestnik evreskogo universiteta v Moskve, Vol 2(18), 1998, 92; Ilya Altman, Victims of Hate – The Holocaust in the USSR 1941-1945 (Moscow, 2002), 443 [Both in Russian].

21. The Jewish Partisans, Vol. 2, 515; Arad, Soviet Union and Annexed Territories, Vol 2, 763-764, 793; Josef Halperin, Youth in Bonds (Tel Aviv, 2002), 97; Bauer, "Kurenets", 79 [All in Hebrew]; Yad Vashem Archive, O.3, file 5763 [In Russian].
22. Arad, "Shoat Yehudey Brit-HaMoatsot", 3 [In Hebrew].
23. Yad Vashem Archive, O.3, file 3712 [In Hebrew]; Rein, "Otnoshenie belorusskogo naseleniia k evreiam", 190; Daniel Romanovsky, "Otnoshenia mezhdu evreiami i okruzhaiushchim neevreiskim naseleniem v vostochnoi Belorussii v period vtoroi mirovoi voiny – k postanovke problemy", Evrei Belorusi – Istoriia i kultura, Vol V (Minsk, 2000), 98 [Both in Russian]; Daniel Romanovsky, "The Holocaust in the Eyes of Homo Sovieticus: A Survey Based on Northeastern Belorussia and Northwestern Russia, in: Holocaust and Genocide Studies, Vol.13 No. 3 (Winter 1999), 367, 371.
24. Cholawski, The Jews in Belorussia, 265-266; Shalom Cholawski, Resistance and Partisan Struggle (Jerusalem – Tel Aviv, 2001), 342; Smolar, Soviet Jews Behind the Ghetto Barrier, 101; Avraham Aviel, Dogalishok (Tel Aviv, 1995), 197; Yad Vashem Archive, O.33c, file 2126 [All in Hebrew]; Yad Vashem Archive, O.3, file 4693 [In Russian]; Orbach, "The Destruction of the Jews", 27.
25. Vinnitsa, "Orshanskoe getto", Evreiskii kamerton, 25.10.2001, 8.
26. Altman, Victims of Hate, 439. In 1932 the name Koidanov was changed to Dzerzhinsk.
27. For more examples see also: The Encyclopedia of the Righteous Among the Nation– Rescuers of Jews during the Holocaust, Supplementary Volumes 2000-2005, Avraham Milgram (ed.) (Jerusalem, 2010), Vol. 1, 7–39.
28. N. Makarov, "KPSS – Organizator vsenarodnoi borby na vremenno okkupirovannoi vragom territorii SSSR", Velikaia pobeda sovetskogo naroda 1941-1945 (Moscow, 1976), 162-163; Nepokorennaia Belorussiia, 21; "Iz vospominanii pervogo sekretaria", 128-129; Iosif Stalin, O Velikoi otechestvennoi voine Sovetskogo Soyuza (Moscow, 1947), 9-17. [All in Russian].
29. Ivan Britov, Partizanskoe dvizhenie na territorii Belorussii v pervyi period Velikoiotechestvennoi voiny (iiul 1941- noiabr 1942) (Minsk, 1997), 4-5; Velikaia otechestvennaia voina 1941-1945 – Entsiklopediia, edited by M. Kozlov et al (Moscow, 1985), 530, 532-533; Viacheslav Boiarskii, Partizany i armiia (Minsk-Noscow, 2001), 69.
30. Cholawski, Resistance and Partisan Struggle, 18-19, 57; Sara Shner-Nishmit, The 51st Brigade (Tel Aviv, 1990), 12-15; Arad, Soviet Union and Annexed Territories, Vol 1, 455; Shalom Cholawski, "Makhteret UPartizanim MiGhetto Kurenets", Yalkut Moreshet 59 (1995), 65; Yad Vashem Archive, O.3, file 1358 [All in Hebrew]; Boiarskii, Partizany i armiia, 106-107; Britov, Partizanskoe dvizhenie na territorii Belorussii, 8; Vsenarodnaia borba v Belorussii protiv nemetsko-fashistskikh zakhvatchikov v gody Velikoi Otechestvennoi Voiny, edited by A. Kuzmin et al (Minsk, 1985), Vol 3, 337; Ilya Altman, Holocaust i evreiskoe soprotivlenie na okkupirovannoi terrotorii SSSR (Moscow, 2002), 229 [All in Russian].

31. Arad, Soviet Union and Annexed Territories, Vol 1, 460; Moshe Kahanovich, The Fighting of the Jewish Partisans in Eastern Europe (Tel Aviv 1954), 35-37 [Both in Hebrew]; Panteleimon Ponomarenko, Vsenarodnaia borba v tylu nemetsko-fashistskikh zakhvatchikov 1941-1944 (Moscow, 1986), 124-127 [In Russian].
32. Yosef Litvak, "Yehudey Brit-HaMoatsot BaMilkhama HaPeila – BaTsava UVaTnua HaPartizanit – BeMilkhemet HaOlam HaShniya", Gesher 2-3(47-48), 1966, 208 [In Hebrew].
33. Boiarskii , Partizany i armiia, 182-183, 188-189; Britov, Partizanskoe dvizhenie na territorii Belorussii, 12-13; Makarov, "KPSS", 166; Rudenkova, "O nekotorykh problemakh", 99; Ponomarenko, Vsenarodnaia borba v tylu, 72-75 [All in Russian].
34. Kahanovich, The Fighting of the Jewish Partisans, 70-73; Avraham Zvi Brown, "HaPartizan HaYehudi BiTnuat HaPartizanim HaSovyetit", Yad Vashem 4 (1948), 175-176; Shalom Yoran, The Defiant – A True Story (Tel Aviv, 1998), 132; Chaim Lazar, "Kakh Nilkhemu HaYehudim", Gesher 1-2 (54-55), March 1968, 88 [All in Hebrew]; Rudenkova, "O nekotorykh problemakh", 100, 102; Vladimir Levin and David Meltser, Chornaia kniga s krasnymi stranitsami (Baltimore, 1996), 520-521; Vitali Perezhogin, "Partizany i naselenie (1941-1945)", Otechestvennaia istoriia No 6 (1997), 153 [All in Russian].
35. Kahanovich, The Fighting of the Jewish Partisans, 45; Yad Vashem Archive, O.3, file 421 [Both in Hebrew]; Britov, Partizanskoe dvizhenie na territorii Belorussii, 13; Stalin, O Velikoii otechestvennoii voiine, 58, 95, 103, 141; Gennadii Volchok, "Partizanskoe dvizheniie v Mogilevskoi oblasti v gody Velikoi Otechestvennoi Voiny (1941-1944)", Vesnyk Mahileuskaha dziarzhaunaha universiteta imia A.A. Kuliashova, Vol 2-3(6), 2000, 167, 170 [All in Russian].
36. Kahanovich, The Fighting of the Jewish Partisans, 45-46.
37. Cholawski, Resistance and Partisan Struggle, 37 [In Hebrew]; Yad Vashem Archive, M.41, files 8 and 34; Vladimir Ermolovich, "Analiz dostovernosti istoricheskogo istochnika", Staronkivaennaii historyi Belarusi Vol 2 (Minsk, 1998), 146-147, 152; Vsenarodnoe Partizanskoe dvizhenie v Belorussii v gody Velikoi Otechestvennoi Voiny–Dokumenty i materialy, edited by Vasilii Shauro et al (Minsk, 1967-1982), 247-248; Vladimir Yakovanko, Na okkupirovannoi zemle (Minsk, 1968), 189; Vasilii Kozlov, Liudi osobogo sklada (Moscow, 1952), 236 [All in Russian].
38. On the fight of the Germans against the partisans see for example: Arad, Soviet Union and Annexed Territories, Vol 1, 349-350, 471; Buchler, "Retsakh HaYehudim BeMasve shel Lekhima", 67-68; Cholawski, The Jews in Belorussia, 75-76 [All in Hebrew]; Yad Vashem, M.41, files 10, 92-93, 112, 256; Emmanuil Ioffe, Belorusskie evrei – tragediia i geroizm 1941-1945 (Minsk, 2003), 261-262 [All in Russian]; Martin Dean, Collaboration in the Holocaust — Crimes of the Local Police in Belorussia and Ukraine, 1941-1944 (York, 2000), 108, 116, 131-132.
39. Brown, "HaPartizan HaYehudi BiTnuat HaPartizanim", 171; HaMaavak LeHaim shel Yehudey Baranovich (Tel Aviv, 1992), 141; The Jewish Partisans, Vol 1, 339 and Vol 2, 619; Kahanovich, The Fighting of the Jewish Partisans, 56-57. [All in Hebrew].

40. Yad Vashem Archive, M.41, files 10, 192; Yad Vashem Archive, O.3, file 6857; Ralph Mavrogordato and Frederick Earl Ziemke, "The Polotsk Lowland", Soviet Partisans in World War II, (ed.), John Armstrong (Wisconsin, 1964), 532; Britov, Partizanskoe dvizhenie na territorii Belorussii, 12 [In Russian].
41. Cholawski, Resistance and Partisan Struggle, 39 [In Hebrew]; Vyacheslav Tamarkin, Eto bylo ne vo sne (Moscow, 1998), 139; Makarov, "KPSS", 169-170; "Iz vospominanii pervogo sekretaria", 37 [Both in Russian].
42. Arad, "Shoat Yehudey Brit-HaMoatsot", 31 [In Hebrew].
43. Brown, "HaPartizan HaYehudi BiTnuat HaPartizanim", 183-184 [In Hebrew]. More examples see: The Jewish Partisans, 167; Solomian-Lutz, Neara Mul Gardom, 103 [Both in Hebrew]; Yad Vashem Archive, O.3, files 3712 [Hebrew], 6071, 6078 [Both in Russian], 7513 [Hebrew], 7579 [Russian], 8445 [Hebrew].
44. In the "Old" Soviet territories – in the 1920s and 1930. In the annexed territories this was done right after the annexation. See for example: Arad, Soviet Union and Annexed Territories, Vol 1, 37-62, 99-141.
45. See for example: Shlomo Kless, "Hashomer Hatsair" under Soviet Regime, September 1939-June 1941 (Tel Aviv, 1999) [In Hebrew].
46. Yitzhak Arad, "HaHitnagdut HaMezuyenet HaYehudit BeShitkheii Brit Hamoatsot HaKvushim – Ghettaot Minsk uVilna", Masuah 21 (1993), 32-33; Halperin, Youth in Bonds, 114; BeKhavleii Klaya – Yehudey Rusia BaShoa HaNatsit, 1941-1943, edited by Benyamin Vest (Tel Aviv, 1963), 20; Mul HaOyev HaNatsi, edited by Smuel Burshtein et al (Tel Aviv, 1961-1986), Vol 3, 22. [All in Hebrew].
47. Kahanovich, The Fighting of the Jewish Partisans, 117; Cholawski, Resistance and Partisan Struggle, 122, 128; HaMaavak LeHaim shel Yehudey Baranovich, 193; With Proud Bearing – Chapters in the History of Jewish Fighting in the Narotch Forests, edited by Moshe Kalchheim (Tel Aviv 1991), 128-129 [All in Hebrew]. In some of the ghettos, the underground planned a revolt inside the ghetto. See for example: Mul HaOyev HaNatsi, Vol 3, 22.
48. Kahanovich, The Fighting of the Jewish Partisans, 96-97; Dov Levin, "Khelkam shel Yehudey Mizrakh Eyropa BaLekhima HaPartizanit BeMilkhemet HaOlam HaShniya", Masuah 13 (1985), 185; Arad, "Shoat Yehudey Brit-HaMoatsot", 34; Sefer Milkhamot HaGetaot – Beyn HaKhomot, BaMakhanot, BaYearot, edited by Yitzhak Zukerman et al (Lohamei Haghetaot, 1954). 479; Yad Vashem Archive, O.3, files 1362 and 3927; Yad Vashem Archive, O.33, file 6350 [All in Hebrew]; Albert Lapidus, "Nas malo ostalos, nam mnogo dostalos", www.vestnik.com/issues/2003/0122/win/lapidus.htm [In Russian].
49. Arkadii Levin, "Evrei i partizany Belorussii v gody natsistskogo genotsida", Ten Kholokosta – Materialy II mezhdunarodnogo simpoziuma Uroki Kholokosta i sovremennaia Rossiia, Moskva 407 maya 1997 (Moscow 1998), 90-91 [In Russian].
50. Tikva Fatal-Knaani, Grodno is not the Same – The Jewish Community in Grodno and Its Vicinity during the Second World War and the Holocaust 1939-1943 (Jerusalem, 2001), 243; With Proud Bearing, 269; Avraham Lidovsky,

BaYearot – Reshimot shel Partizan Yehudi (Tel Aviv, 1946), 59 [Both in Hebrew]; Lapidus, "Nas malo ostalos" [In Russian]; Romanovsky, "Soviet Jews under Nazi Occupation", 246; Isaac Kowalski, Anthology on Resistance 1939-1945 (USA, 1984), Vol.I, 113.

51. Testimony to Revolt and Destruction, edited by Chaya and Chaim Lazar (Tel Aviv, 1978), 31 [In Hebrew]; Lapidus, "Nas malo ostalos" [In Russian]; Yad Vashem Archive, M.41, file 12.
52. Bobruisk Memorial Book, 735-736 [In Hebrew]; Smilovitsky, "Righteous Gentiles, the Partisans and Jewish Survival", 307.
53. Arad, "HaHitnagdut HaMezuyenet HaYehudit", 45 [In Hebrew]; Smilovitsky, Yevrei Belarusi, 146-147 [In Russian].
54. Shner-Nishmit, The 51st Brigade, 11; Cholawski, Resistance and Partisan Struggle, 248, 350; The Jewish Partisans, Vol 1, 166, 190; Levin, "Khelkam shel Yehudey Mizrakh Eyropa", 149 [All in Hebrew]; Yad Vashem Archive. O.33, file 151; Ioffe, Belorusskie evrei, 307 [In Russian].
55. Levin, "Khelkam shel Yehudey Mizrakh Eyropa", 150; Mordechai Marmurshtein, "Al Na Tomar Hine Darki HaAkhrona", Zeraim (December 1969); Gad Rosenblat, "Roman Plaska – Partizan VeKhayal", Kol Nekhey Milkhama, Vol 150, 31 [Both in Hebrew]; Smilovitsky, "Righteous Gentiles, the Partisans and Jewish Survival", 305; Adam Rogovski, Lost and Found (Tel-Aviv, [n. d.]), 115.
56. Lidovsky, BaYearot, 81 [In Hebrew].
57. Yad Vashem Archive, M.41, file 10; Smilovitsky, "Righteous Gentiles, the Partisans and Jewish Survival", 305; Rein, "Otnoshenie belorusskogo naseleniia k evreiam", 197-198; Shapiro, "Kopel Kolpanitskii", 22 [Both in Russian].
58. Berestizki, Run for Your Life, 141 [In Hebrew].
59. Litvak, "Yehudey Brit-HaMoatsot BaMilkhama HaPeila", 208-209 [In Hebrew]; Iosif Maliar, Evreiskii narod v borbe protiv natsizma v gody Vtoroi mirovoi voiny (1939-1945) (Lohamei Haghetaot), 57-58 [In Russian].
60. Slepyan, "The Soviet Partisan Movement and the Holocaust", 7.
61. Cholawski, "Minsk BeMaavaka VeKilyona", 102; Arad, "HaHitnagdut HaMezuyenet HaYehudit", pp. 33-34 [Both in Hebrew]; Lavon Institute for Labour Research, Benyamin Vest Collection, IV-104-53, file 63; Yuri Suhl, "The Resistance Movement in the Ghetto of Minsk", in: Yuri Suhl (ed.), They Fought Back: The Story of the Jewish Resistance in Nazi Europe (NY, 1975), 232; Albert Maizel, "O neizvestnom pisme G.Smoliara Minskomu podpolnomu komitetu KP(b)B", Evrei Belarusi – Istoriia i Kultura, Vol II (Minsk, 1998), 193 [In Russian]. Not in all Ghettos the Judenrat collaborated with the underground. See for example: Sara Shner-Nishmit, VeEl HaMenukha Lo Bati (Lohamei Haghetaot, 1986), 102 [In Hebrew].
62. Dina Porat, "Khelkam shel HaKomunistim HaYehudim BaMakhtarot, BaGetaot UVaYetsia LaYearot – Shalosh Dugmaot", Dapim LeKheker Tkufat HaShoa, Vol 12 (1994), 125-126 [In Hebrew]. The printing equipment was later transferred from the ghetto to the woods. See: Yaakov Grinshtein, Ud MiKikar HaYovel

(Tel Aviv, 1967), p. 55 [In Hebrew].
63. Suhl, "The Resistance Movement in the Ghetto of Minsk", 234.
64. Ibid, 235; The Jewish Partisans, Vol 1, 514; Minsk, Jewish Mother-City, Vol. 2, 290 [Both in Hebrew]. The author and painter Moshe Levin was known in the Minsk Ghetto for his ability to forge documents. See: ibid, 289.
65. Arad, "Shoat Yehudey Brit-HaMoatsot", 33-34; Arad, "HaHitnagdut HaMezuyenet HaYehudit", 43-44 [Both in Hebrew].
66. Ibid, 35-36; Minsk, Jewish Mother-City, Vol 2, 287-289.
67. Arad, "HaHitnagdut HaMezuyenet HaYehudit", 36-37; Minsk, Jewish Mother-City, Vol 2, 289-290; Smolar, Soviet Jews Behind the Ghetto Barrier, 100; Kahanovich, The Fighting of the Jewish Partisans, 93 [All in Hebrew]; Suhl, "The Resistance Movement in the Ghetto of Minsk", 234.
68. Cholawski, Resistance and Partisan Struggle, 353, 360 [In Hebrew].
69. Kahanovich, The Fighting of the Jewish Partisans, 135-137 [In Hebrew]; Yad Vashem Archive, O.3, files 1358, 3955; Leonid Smilovitsky, Katastrofa evreev v Belorusii 1941-1944 (Tel Aviv, 2000), 134; Ioffe, Belorusskie evrei, p. 260 [Both in Russian].
70. Brown, "HaPartizan HaYehudi BiTnuat HaPartizanim", 180 [In Hebrew]; National Archive of the Republic of Belarus, 3500-1, file 163, 80; Ilya Altman, Kholokost i evreiskoesoprotivlenie (Moscow, 2002), 253-254 [In Russian].
71. Biographical Dictionary of Jewish Resistance, edited by Yehiel Granatstein and Moshe Kahanovich (Jerusalem 1965), Vol 1a, 119; Ibid, Vol 1b, 48 [In Hebrew].
72. The Jewish Partisans, Vol 1, 530 [In Hebrew].
73. Yad Vashem Archive, M.41, file 177.
74. National Archive of the Republic of Belarus, 3500-1, file 161, 78; Ibid, file 162, 26; Ibid, file 163, 124-125; Altman, Victims of Hate, 360 [In Russian].
75. Yad Vashem Archive, O.3, file 4225; Gad Rosenblat, "Partizany", Slovo invalida voiny, Vol8, 36 [In Russian].
76. Kahanovich, The Fighting of the Jewish Partisans, 213-214; Brown, "HaPartizan HaYehudi BiTnuat HaPartizanim", 178; Litvak, "Yehudey Brit-HaMoatsot BaMilkhama HaPeila", p. 215; Biographical Dictionary of Jewish Resistance, Vol 1a, 115-116 [All in Hebrew]; Sovetskiye evrei pishut Ilie Erenburgu 1943-1966, edited by Mordechai Altshuler et al (Jerusalem, 1993), 59 [In Russian]; Dean, Collaboration in the Holocaust, 139.
77. The Jewish Partisans, Vol 1, 402; Kaplan, Ad Yon Kipurim Ze, 78, 85; Kahanovich, The Fighting of the Jewish Partisans, 219, 221-222; Yad Vashem Archive, O.3, file 3188.
78. Kahanovich, The Fighting of the Jewish Partisans, 316-317; Brown, "HaPartizan HaYehudi BiTnuat HaPartizanim", 178; Yad Vashem Archive, O.3, files 3926, 6333.
79. With Proud Bearing, 225; Kahanovich, The Fighting of the Jewish Partisans, 318-319; Yad Vashem Archive, O.3, file 7729.
80. Kahanovich, The Fighting of the Jewish Partisans, 211, 365-381; Yad Vashem

Archive, O.33, file 151, Ibid, O.3, file 4890; Lazar, "Kakh Nilkhemu HaYehudim", p. 89.

81. Moshe Tsimkind, "Odin iz diversionnoi gruppy", Evreiskii kamerton, 2.11.2000, 6 [In Russian].
82. Andreii Plotkin, Podvigov ne sovershal (Moscow, 2000), 66 [In Russian].
83. About what was known to the Soviet partisan movement see: Yaakov Falkov, "Kemo BeKhol Makom, HaYehudim Mushmadim KeAm – Retsakh Yehudey Brith HaMoatsot BeDivukhey HaModiin shel HaPartizanim HaSovyetim", Dapim LeKheker HaShoa 22, 239-264 [In Hebrew].
84. Cholawski, Resistance and Partisan Struggle, 47 [In Hebrew]; Evgenii Rozenblat, Irina Elenskaia, Pinskie evrei 1939-1944 (Brest, 1997), 191; David Meltser, "Neizvestnaia stranitsa evreiskogo soprotivleniia", Vestnik, No 7(240), 29 March 2000, www.vestnik.com/issues/2000/0329/win/meltser.htm; Smilovitsky, "Righteous Gentiles, the Partisans and Jewish Survival", 312-313; David Meltser, "Evreiskoe antinatsistskoe soprotivlenie", Vestnik, No 14(221), 6 July 1999, www.vestnik.com/issues/1999/0706/koi/meltzer.htm; Arkadii Leizerov, "Dvulitaia pravda", Mishpokha, No 3, 1997, 79; Boris Sokolov, Okkupatsiia – Pravda i mify, http://militera.lib.ru/research/sokolov3;Nepokorennaia Belorussia, 70 [All in Russian]; Slepyan, "The Soviet Partisan Movement and the Holocaust", 11-12.
85. Arad, Soviet Union and Annexed Territories, Vol 2, 96 1[In Hebrew]; Altman, Victims of Hate, 399-400 [In Russian].
86. Meltser, "Neizvestnaia stranitsa evreiskogo soprotivleniia", 86 [In Russian].
87. Smilovitsky, Katastrofa evreev v Belorusii, 149; Galburt, Vospominania, 35; Abram Rubenchik, Pravda o Minskom getto (Tel Aviv, 1999), 94 [In Russian].
88. In the summer of 1943 the Polish government in exile established a partisan movement, Armia Krajowa (national army). Most of the Poles deserted the Soviet units and joined AK. Even though its aim was to fight against the Germans, it also fought against the Soviets and did not accept Jews into its ranks. See for example: Yoran, The Defiant, 161 [In Hebrew].
89. Arad, "Shoat Yehudey Brit-HaMoatsot", p. 35; Kaplan, Ad Yon Kipurim Ze, 51-52; Smilovitsky, "Yehudim UPolanim BaTnua HaPartizanit", 86; HaMaavak LeHaim, 207 [All in Hebrew]; Yad Vashem Archive, O.3, file 4890; Levin, "Evrei i partizany Belorussii, 89 [In Russian].
90. Avraham Keren-Paz, "Derekh HaYesurim shel Partizan Yehudi", Pirsumey Muzeon HaLokhamim veHaPartizanim, Vol 5(20), 1973, 20 [In Hebrew]; Yad Vashem Archive, O.3, file 5053; Yakovanko, Na okkupirovannoi zemle, 36-37 [In Russian]; Slepyan, "The Soviet Partisan Movement and the Holocaust", 14-15.
91. Biographical Dictionary of Jewish Resistance, Vol 1a, 42-43; Lidovsky, BaYearot, 119-120 [Both in Hebrew].
92. Kahanovich, The Fighting of the Jewish Partisans, 183-187; Lazar, "Kakh Nilkhemu HaYehudim", 87; Solomian-Lutz, Neara Mul Gardom, 106 [All in Hebrew].
93. Kahanovich, The Fighting of the Jewish Partisans, 187-193; Litvak, "Yehudey Brit-HaMoatsot BaMilkhama HaPeila", 208-210; Kaplan, Ad Yon Kipurim Ze, 58, 79;

Shner-Nishmit, The 51st Brigade, 15 [All in Hebrew]; Lapidus, "Nas malo ostalos"; Rein, "Otnoshenie belorusskogo naseleniia k evreiam", 195 [Both in Russian].

94. Kahanovich, The Fighting of the Jewish Partisans, 200, 357-358, 360; Cholawski, Resistance and Partisan Struggle, 314; Eliezer Lidovsky, VeShviv HaEsh lo Daakh (Tel Aviv, 1986), 138; Levin, "Khelkam shel Yehudey Mizrakh Eyropa", 149 [All in Hebrew].
95. Kaplan, Ad Yon Kipurim Ze, 113 [In Hebrew].
96. With Proud Bearing, 246-247 [In Hebrew]; Yad Vashem Archive, O.3, file 421.
97. Arad, "Shoat Yehudey Brit-HaMoatsot", 35; Kahanovich, The Fighting of the Jewish Partisans, 138-139, 173-174; Testimony to Revolt and Destruction, 59; With Proud Bearing, 305; The Jewish Partisans, Vol 1, 177; Marmurshtein, "Al Na Tomar" [All in Hebrew]; Slepyan, "The Soviet Partisan Movement and the Holocaust", 15; Yad Vashem Archive, M.41, file 123.
98. Kahanovich, The Fighting of the Jewish Partisans, 195-197 [In Hebrew]; Yad Vashem, M.41, file 2450; Yad Vashem Archive, O.3, files 3701, 3753; Levin, "Evrei i partizany Belorussii", 91; Meltser, "Evreiskoe antinatsistskoe soprotivlenie"; Naum Pravdivyii, "Belorusskie partizany ubivali evreev – o chom molchali partizany v belorusskikh lesakh", www.jewish.ru/994158100.asp [In Russian]; Slepyan, "The Soviet Partisan Movement and the Holocaust", 5.
99. Kopel Kolpanitskii, Nigzar LeKhaim – Sipuro shel Nitsol Ghetto Lakhva (Tel Aviv, 1999), 70; Yoran, The Defiant, 137; David IoKhevidovich Kahani, Gilgul Mekhilot (Tel Aviv, 1973), 74; Biographical Dictionary of Jewish Resistance, Vol 1b, 58-59; Litvak, "Yehudey Brit-HaMoatsot BaMilkhama HaPeila", 211; With Proud Bearing, 19-20, 305; The Jewish Partisans, Vol 1, 608; Keren-Paz, "Derekh HaYesurim", 12; Lidovsky, BaYearot, 103 [All in Hebrew]; Yad Vashem Archive, O.3, file 3081; Shapiro, "Kopel Kolpanitskii", 22; Avraam Ingber, "Vo ves rost", Slovo invalida voiny, No 5, 58 [Both in Russian]; Faye Schulman, A Partisan's Memoir (Toronto, 1995), 109.
100. Kahanovich, The Fighting of the Jewish Partisans, 174, 176; Cholawski, Resistance and Partisan Struggle, p. 60-68; Kolpanitskii, Nigzar LeKhaim, 111; Moshe Shutan, Steps in the Fog (Tel Aviv, 1996), 57 [All in Hebrew]; Yad Vashem Archive, O.33, file 631; Maizel, "O neizvestnom pisme G.Smoliara, 194; Lapidus, "Nas malo ostalos" [Both in Russian].
101. Arad, "HaHitnagdut HaMezuyenet HaYehudit", 47; Levin, "Khelkam shel Yehudey Mizrakh Eyropa", 150-151; Lazar, "Kakh Nilkhemu HaYehudim", 87; Moshe Kahanovich, "Mipneii ma lo Kama Tnua Nifredet shel Partizanim Yehudim BeMilkhemet HaOlam HaShniya?" Yad Vashem, Vol 1 (1957), 112; The Jewish Partisans, Vol 1, 528 [All in Hebrew]; Yakov Shepetinskii, Prigovor (Tel Aviv, 2002), 57 [In Russian].
102. There are countless examples to this. See for instance: Aviel, Dogalishok, 186; With Proud Bearing, 228; Cholawski, Resistance and Partisan Struggle, 367-369; Mul HaOyev HaNatsi, Vol 1, 69-73; Solomian-Lutz, Neara Mul Gardom, 100;

Shner-Nishmit, VeEl HaMenukha, 145, 149; The Jewish Partisans, Vol 1, 233 [All in Hebrew]; Yad Vashem Archive, M. 41, file 10; Yad Vashem Archive, O.3, files 1358, 3895, 4185, 5915, 8272; Altman, Holocaust i evreiskoe soprotivlenie, 233; Rubenchik, Pravda o Minskom getto, p. 24 [Both in Russian]; Schulman, A Partisan's Memoir, 96–97, 143–144.

103. Berestizki, Run for Your Life, 127 [In Hebrew]
104. Brown, "HaPartizan HaYehudi BiTnuat HaPartizanim", 187; Biographical Dictionary of Jewish Resistance, Vol 1a, 112; Ibid, Vol 1b, 19; Lidovsky, VeShviv HaEsh lo Daakh, 105-106; Aviel, Dogalishok, 191-193; Levin, "Khelkam shel Yehudey Mizrakh Eyropa", 150 [All in Hebrew]; Altman, Victims of Hate, 371.
105. Shutan, Steps in the Fog, 26.
106. Lidovsky, VeShviv HaEsh lo Daakh, 137-138; Halperin, Youth in Bonds, 117; Leonid Smilovitsky, "Antishemiyut BaPartizanka HaByelorusit", Yalkut Moreshet 59 (1995), 53-54; Gad Rozenblat, "Partizanim Yehudim BaTnua HaPartizanit HaSovyetit", Kol Nekhey Milkhama 153, 28 [All in Hebrew]; Yad Vashem Archive, M.41, file 43; Ioffe, Belorusskie evrei, 317 [In Russian]; Schulman, A Partisan's Memoir, 102.
107. Levin and Meltser, Chornaya kniga s krasnymi stranitsami, 457-458 [In Russian].
108. Smilovitsky, "Yehudim UPolanim", 87 [In Hebrew]; Smilovitsky, "Righteous Gentiles,the Partisans and Jewish Survival", 313; Solomon Shvarts, Antisemitizm v Sovetskom Soiuze (New York, 1952), 158-159; Unichtozhenie evreev SSSR v gody nemetskoi okkupatsii (1941-1944), edited by Yitzhak Arad (Jerusalem, 1992), 381 [Both in Russian].
109. Sergeii Berkner, "Evrei v borbe s nemetskim fashizmom v tylu vraga", Korni, No 10 (July-August 1998), www.shorashim.narod.ru/Berk.htm. [In Russian].
110. "Great Land" – the name which the partisans gave to the hinterland.
111. A. AndL. Leizerov, "Cherezliniiufronta", Aviv, 7/2000, 11; Arkadii Leizerov, "Unikalnyisluchaiiistorii Velikoiotechestvenoiivoiny", Betsavta 11 (2000), 4 [Both in Russian].
112. Minsk, Jewish Mother-City, Vol 2, 292 [In Hebrew]; Vsenarodnaia borba v Belorussii, 331 [In Russian].
113. Testimony to Revolt and Destruction, 13 [In Hebrew].
114. Shner-Nishmit, The 51st Brigade, 35 [In Hebrew].

Translated by Miriam Shimoni

Graciela Ben-Dror

"Traitors and Evildoers": Antisemitic Stereotypes in Catholic Textbooks in Spain under Franco

Introduction

During 1942, and definitely by the end of the year, the world knew about the systematic, mass killing of the Jews of Europe. The reports that appeared on the subject throughout the year were officially confirmed by Rabbi Stephen Wise at a news conference held in November 1942, and one month later in declarations issued by the governments of the Allied Powers. At the time, these reports may not have been fully appreciated and did not garner broad public resonance throughout Europe. Nevertheless, during the war, and certainly by its conclusion, the fact that a Holocaust had befallen the Jewish people was general knowledge. This raises a number of compelling questions regarding the impact of this knowledge on attitudes toward the Jews in Europe during and immediately following the war. What were the reactions to these reports and to the awareness they engendered? Did Antisemitic views continue to prevail, or did the awareness of the tragic events result in changes in opinion or expressions of increased benevolence?

This article explores the reaction in Spain, under the rule of General Francisco Franco, to the wartime public knowledge of the ongoing murder of European Jewry. Numerous studies have considered Spain's policies toward the Jews during WWII regarding the rescue of Jews, the issue of entry and transit permits, the provision of consular protection and repatriation. Most conclude that, in light of the great potential it possessed as a neutral country that had declared non-belligerence, Spain's efforts to save Jews during the war were inadequate to its potential possibilities.[1]

This article examines a different aspect of Spanish policy and practice during and following the Holocaust: the realm of education. It focuses on the educational methods and material employed in Spain, from Franco's rise to power in 1939 until his death in the mid-1970s, to achieve the country's educational goals vis-à-vis the Jews. In order to better understand this issue, this article also considers the regime's ideological and political background and the general educational content that left its impact on the Jewish question.

The 1939 victory of the Nationalists over the Republican Forces in the bloody Spanish Civil War of 1936-1939 paved the way for the institution of Franco's autocratic, authoritarian regime following WWII and the victory of the Allied Powers. Franco's coalition, which was based on the forces that supported the Nationalist war effort during the Civil War, comprised representation of different and, at times, contradictory ideological and political interests. The nationalist coalition by means of which Franco ruled rested on the army and a number of political families: the fascist *Falange Española* party, monarchists of all streams, and the Catholic Church. After the Nationalist victory, each of these forces was allocated a share of the positions within the regime. Before the war even ended, the Falange party and the army enjoyed a unique status of prestige and held senior positions. This state of affairs continued as long as the Axis Powers retained the upper hand during the war.

In order to ensure the survival of his regime in the new international context that emerged after WWII, Franco attempted to bring himself closer to the countries of the West. One measure he took to this end was to define his regime as National-Catholic (*nacional-catolicismo*), a step aimed at shaking off his regime's widespread image in the West as a fascist regime and at gaining acceptance into the family of victorious nations. Franco's regime also benefited from the legitimacy that stemmed from the decision of the Catholic political leaders in Spain to join his government. In return, the Church was awarded positions of major strategic and political value with regard to its own interests.[2]

The official Catholic Church sought to appropriate positions that would enable it to control religious and social aspects of public life, moral values, and the education of the masses. With this goal in mind, it also requested and received control of the Spanish Ministry of Education, which it maintained over the years. The Church supervised the education system and the religious and moral education of faithful Catholics in both private and state schools, and was involved in all aspects of the work. According to Spanish historian Julián Casanova

> The Church became the true soul of Franco's new state when it came back to life after the blow it sustained from the anti-clerics. The Church and the Catholic religion now flooded education, customs, the administration, and centers of power.[3]

After coming to power, Franco took a number of measures against the Republican camp, which emerged defeated from the Civil War and was viewed as a symbol of betrayal of the homeland and an enemy of

Spain. On an ideological level, conceptions of the Civil War as a "sacred war" and the members of the Republican camp as "traitors" were applied not only to the Freemasons and Liberal Communists, but also to the Jews. Indeed, after the Civil War, the regime's goal vis-à-vis the Jews was to reduce and restrict their freedom of action.[4]

Between March and October 1940, prohibitions were instituted on Jewish marriage ceremonies, Jewish burials and Jewish ritual circumcisions, and Jewish babies could not be registered in the population registry without first being baptized. In Barcelona, the Jewish cemetery was destroyed and all Jewish institutions were disbanded. In this context, many Jewish families who had fled Nazi Germany to Spain and sought to continue living in the country now converted to Catholicism. Jewish children studying in schools of the state education system were required to attend classes on the Catholic religion ("Catechism" classes), and to use special text books for subjects related to religion (such as ethics and religious history), which were written specifically for children and teenagers.[5]

During this period, the Church establishment neither intervened in Jewish affairs nor issued any official declarations or documents regarding the Jews, with the exception of an anti-Jewish pastoral letter issued by Carmelo Ballester y Nieto, the bishop of León. The Church's position on the Jews at the time was reflected in short oral statements or brief phrases included in pastoral letters issued by bishops, journals published by the Church leadership and the important religious orders, official Catholic periodicals and newspapers, books written by senior clergymen, and other Church sources.[6]

Within this historical context, this article examines the educational policies of the Franco regime and the Catholic Church in Spain with regard to the Jews. It seeks to determine whether receiving reports regarding the Jewish genocide carried out by the Nazis and their allies had an impact (i.e. whether these reports sparked a change in the traditional Antisemitic views prevalent among the Spanish Catholics within official church frameworks and the Franco government.) This article also attempts to clarify whether any changes in position took place later, in the wake of the Second Vatican Council of 1965, which changed the Church's position on the Jews with its "Declaration on the Relation of the Church to non-Christian Religions," the "*Nostra Aetate*" ("In our Age").[7] Examining the Catholic Church from such a local-national perspective adds a new layer to the scholarship on the Antisemitism that emerged among Catholics in general, and Catholics in Spain in particular, between the two world wars, during WWII, and thereafter.

This article is based on a survey of the textbooks written by the Catholic state education system and approved by the Church censorship and the local archbishop. It focuses on the Antisemitic contents of these books, asks whether this content was maintained after the Holocaust, and considers the nature and timing of subsequent changes. In order to understand the manner in which the Franco regime dealt with the subject of the Jews in the realm of education it is first necessary to consider the issue in the context of the regime's overall worldview, the values it promoted, and the education with which it sought to provide the children of Spain, as reflected in the textbooks designated for children in the state and private Catholic education system. Against this background, this article examines the distinct educational message regarding the Jews articulated in the textbooks during the period in question.

The Ideological Context in Textbooks: Values, Worldview, and Methodology

Textbooks designated for elementary and high school students constituted a major tool of indoctrination in the territories under Nationalist control during the Spanish Civil War, and its importance increased after Franco's victory over the Republican forces in April 1939. The textbooks served as a means of conveying the messages which the political and cultural elite in Spain wished to inculcate in the young minds of the Nationalist state after its victory over the Republicans.

An examination undertaken of a large number of textbooks of this kind provides a clear picture of the educational messages they promoted. Despite the lack of data regarding the number of copies examined and their scope of distribution, the history books nonetheless stand out as one of the primary instruments employed by the regime to shape national identity. The image that emerges is one of history books that placed less of an emphasis on historical fact than on interpretation and moral judgment. The books– which were selective in both the topics they addressed and failed to address, as well as in the specific content chosen for publication– covered only specific periods in Spanish history for ideological reasons.

The books dealt extensively with national identity, the fate of the Spanish nation and the relationship between Spain's past and present, and employed an ideological bias that promoted the worldview favored by their authors and that overshadowed chronological, thematic, and lack historical continuity. The overall goal of this approach was to create an emotional image based on the values of national identity espoused by the authoritarian regime.

The textbooks praised the values of the Spanish Empire during its days of greatness and glorified Catholicism as both a religion and an overall worldview. The regime, which after the defeat of the Republicans sought obsessively to control the ideology of the general population and to suppress liberal democratic anti-clerical worldviews, used these textbooks as a means of indoctrination.

Control over the nation's textbooks offered a rare opportunity to disseminate content that glorified the ruling regime, thereby creating an effective means to achieve ideological control over students and teachers alike. The books served to convey "traditional Spanish pedagogy" to teachers, who in any case had received only the most elementary training. Their main message was "Spain: one, great, and free," or, in other words, a Spain that was traditional, imperial, and Catholic, and very frequently used the term *Hispanidad*, a term used to express a Spanish identity between the Spanish Empire after 1942 and Catholicism. In 1937, short courses were introduced in Nationalist territory to instruct teachers in the values of Nationalist Catholic Spain as opposed to the values of liberal democratic Spain.[8]

The regime's use of education as an effective tool for increasing popular support of Franco also involved portraying Franco and his regime as the heirs of the traditional values of Spanish history. Its educational doctrine focused on the two central values of the new regime: patriotism and religiosity. The names of schools that were not consistent with the new spirit were changed, Republican teachers were dismissed, and anyone who attempted to challenge Catholicism's central role in education was forced out of the system. The Primary Education Act of 1945 defined the main aim of education for the primary grades as forming "the will, conscience, and character of the child toward the fulfillment of his duty and his eternal destiny," and imbuing him with a consciousness of service to the state in accordance with the principles of the fascist Falange movement. The subjects of study used by the system to achieve this goal included Catholic religion, Spanish language, geography and history. Once a week, children were required to attend a class on "political education" conducted by the Fascist Falange Española movement. As emphasized by Education Minister José Ibañez-Martín: "just as our regime is Catholic, the law, too, is Catholic law."[9]

Upon the conclusion of WWII and the victory of the Allied Powers, the regime stressed its Catholic dimension on the educational level and played down the role of its fascist ideology. As we have seen, the regime defined itself as "National-Catholic" and placed an emphasis on patriotic values and ideals and on Spanish sentiment and will. Its aim was to strengthen a sense of patriotic unity that was portrayed as the opposite of the pedagogy of the enemy camp, which was based, according to them, on rationalism, foreignness, and "cold intellectualism."[10]

The content of elementary school textbooks ranged from concepts rooted in Spanish tradition, presented as a collection of absolute facts, to emotional and political expressions regarding historical events. In this way, they sought to immunize the masses against two "viruses" which the regime fully opposed: communism and separatism (*separatismo*). In the lexicon of the regime, "communism" was synonymous with the values of the vanquished Republic, and "separatism" was associated with the demands for Catalan independence and the Basque separatists, who sought independence for the Basque Country. Communism and separatism were defined as hostile concepts that ran contrary to the values of traditional, united, Catholic Spain according to which Franco's regime operated. The history of the Inquisition and the Spanish expulsion of the Muslims and Jews, which will be addressed later in this article, were regarded as subjects whose study was essential to the prevention of crime, divisiveness, and heresy. As previously observed, the textbooks made no mention of most of contemporary Spanish history, jumping from the wars of the integrist Catholic Carlistas, who claimed the right to inherit the crown from the nineteenth century onward, directly to Franco's regime, which was portrayed as a savior delivering the country from the evils of liberalism and communism. The books said almost nothing on the subject of the *other* Spain– that is, the liberal and socialist Spain of the twentieth century.

The same distortions were implanted in textbooks written for high school instruction and preparation for the matriculation exams. Although these books differed somewhat from the elementary school books in the quantity of historical information they contained and their quality of writing, for the most part they too focused primarily on content that was selected for the purpose of glorifying the Franco regime.

One well known book of this kind was the *Manual of the History of Spain* by Catholic right wing intellectual José María Pemán.[11] Pemán's *Manual* was replete with moral and theological judgment. It also contained concepts, terms, and ideas developed by the author himself, which soon began to appear in numerous other textbooks of the time. Examples include his call to continue the struggle to expel alien ideas, his support for the universalization of the truth and of Catholicism (the true religion), and his demand to expel all foreigners, traitors, and "bad" Spaniards. Pemán praised monarchist and Catholic Nationalist Spain, including the Inquisition, and preached against heretics, liberalism and communism. The book also praised Francisco Franco, the country's "*caudillo*" and savior.

Other books highlighted Spain's imperial mission. This emphasis was particularly prominent during the period immediately following the victory over the Republicans, when the Falange party had the upper hand. During this period, the Falangists and the Catholics competed with each other for ideological hegemony within the Education Ministry. During teaching conventions, *Imperial Glory* by Falangist Luis Ortiz Muños was the most widely quoted book of all. The message it imparted was that Spain bore the mission of a great empire and that the Spanish race had led Spain to greatness at various points in history, as reflected in the alliance between Catholic Spain and the Papacy. In this book, as in others, Spain's enemies were primarily internal heretics threatening political unity and

Catholic religious unity. To be true to their history, the children of Spain must defend the Catholic Church, obey the *caudillo*, and live at peace with their Spanish brethren.[12] Another well known book, entitled *I am a Spaniard*, glorified values such as honesty, diligence and kindness, as well as patriotism, Spanish tradition and the values of "sentiment, courage, and good deeds."[13]

At the same time, when tension emerged between values of the Falange and those of the Catholic Church, efforts were made to reach a compromise. Although private Catholic schools did not cultivate values associated with political totalitarianism and adhered primarily to religious Catholic values, the efforts at compromise introduced such values as "Christian totalitarianism," as distinct from the "totalitarianism of the state." This ideological tension continued to exist in the years to come.

On a procedural level, only the textbooks designated for elementary education required government approval to be printed. High school text books did not require such authorization and were written by university professors from the "Professors' Institute," who not only wrote but also supervised the content of high school textbooks and the formulation of questions for the matriculation exams.

Many of the elementary school textbooks published during the 1940s were referred to as *Encyclopedias* and were circulated by publishers belonging to orders of the Catholic Church, such as Bruño of the Salesians (*Salecianos*), Eldevives of the Marist Brothers (*Hermanos Maristas*), and Textos A.P. of the Piarist Fathers (*Padres Piaristas*). Their pedagogical approach was based on methods of the Catechism of the Catholic Church's educational system, such as the memorization of pre-formulated, easy questions and answers.

International developments that took place during the 1950s and that had an impact on the Spanish education system were also reflected in the textbooks. Such developments included the effort to obscure the ideological contradictions manifested in the Spanish Civil War, the rising influence of the United States, the ever intensifying international contacts among European intellectuals, the liberalization of Spanish censorship in the 1960s, and the reforms of the Roman Catholic Church instituted by the Second Vatican Council of 1962-1965. In 1965, the Ministry of Education was renamed "the Ministry of Education and Sciences," indicating a new way of thinking and a new degree of openness. Taking into accoutnt this ideological background of the dissemination of National Catholic values and the National Catholic worldview in Francoist Spain following WWII that we now turn to an analysis of the textbooks' treatment of the Jews and Judaism during the period.[14]

"The Jews" and "Judaism" in Spanish Textbooks

The textbooks dealing with the "history of Christianity," "Christian dogma," and "Christian ethics" –which became mandatory subjects of study in state and private schools under the Franco regime– cast the Jews in a distinctively negative light. Their portrayal was based on traditional and modern Antisemitic stereotypes, but also added Spanish specific motifs.

I. The Jewish People as a "Deicidal People"

One subject discussed at length in the textbooks written for the government education system and the private Catholic education system was the crucifixion of Jesus, in which the Jews are cast as playing a central role. In the narrative advanced by the textbooks, Jews are typically portrayed as being directly responsible for Jesus' death and are consistently referred to as "*the* Jews," making a generalization of all Jews. The books never mention the fact that Jesus, his mother, and his close companions were also Jews. In the version of the story contained in the textbooks, guilt is levelled squarely against the Jews, whose pleas and demands resulted in Jesus' death sentence. The Jews were the "wicked," the "traitors," "the killers of God." The Romans, in contrast, were innocent and merciful, albeit also obedient, in that they consented to the demands of the Jews. Pontius Pilate, the Roman governor of Jerusalem, is typically portrayed as devoid of all authority and, sometimes even as a merciful man who saw no harm in Jesus' words. He is portrayed as having sought to save Jesus' life but as ultimately giving in to the pressure of the Jews. This role reversal, in which the Roman ruler is cast as weak and compassionate and the Jews are portrayed as evil to the core, conveyed a clear message of hatred toward the Jewish people. The illustrations of the crucifixion that appeared alongside the texts were aimed at reinforcing this anti-Jewish content and arousing the emotional revulsion of the young Spanish readers.

The children were provided with a detailed account of the stages of the murder, Jesus' attachment to the cross, and the insertion of nails into his hands and feet. In addition to the illustrations, the books also contained verbal accounts that spoke of a slow and extremely cruel death, from which the Jews benefited most of all. The illustrations, which were designed especially for children, made the story more tangible and were meant to arouse a sense of compassion for Jesus on the cross on the one hand, and hatred for the Jews on the other hand. The story conveyed to the children reflected a sense of complete harmony between Jesus and the Roman ruler and was devoid of historical context regarding the methods of rule of the Roman Empire in the countries it conquered. In this manner, the textbook on "sacred history" that was authorized by the Church authorities stressed the legendary myth of the Jews' responsibility for the murder of God.[15]

Pilate also appears in the textbooks on "Catechism" (*Catequismo*) – religious studies for young children. One such book titled *I Believe in Jesus* included the following account: "In order to satisfy the Jews,

Pilate sought to place a wreath of thorns on his [Jesus'] head, but the Jews demanded his crucifixion." According to this account, Pilate was not evil but rather had a weak personality and was controlled by the Jews. It was the Jews who demanded Jesus' crucifixion and it was they who held decisive influence. Pilate, on the other hand, was convinced of Jesus' innocence and is portrayed as a weak coward who feared the Jews' power and gave in to their demands. The Jews emerge from the narrative advanced in the textbooks as omnipotent, not dissimilar from the image of the Jew in modern Antisemitism. This book, written by a Catholic clergyman, was also published with the authorization of the Church, as granted by the Archbishop of Barcelona and the Church censors.[16]

Varying in verbal formulation but identical in content, the story of the crucifixion of Jesus is repeated in another book titled *Sacred History*, which was designated for the youngest students of elementary school. Here too, Pilate is portrayed as acting out of fear of the Jews and only in in this context agreed "to crucify the Messiah."[17] Another book entitled *Lessons of Interest*, also designated for seven-to-nine year olds, describes the Jews as "the enemies of Jesus" and "the ones who determined his death sentence." During the crucifixion, the Romans –who were the rulers of the land at the time– are conspicuously absent from the narrative and therefore play no role in Jesus' sentence. Only the Jews are present, and therefore they alone are guilty of Jesus' death.[18]

Descriptions of the Jews as "wicked" appear primarily in the textbooks designated for the youngest elementary school students, six-to-seven year olds during their first year of religious studies. To make these books suitable for such a young age, the authors forewent the historical framework entirely and simplified the narrative to produce a distorted historical narrative. The books designated for these young children ~~students~~ repeat the same account, in which the Jews were the ones who demanded Jesus' death and Pontius Pilate was simply a weak man who gave in to their pressure. This message emerges clearly in the following passage from the textbook *We Saw Our Lord*: "The Jews, who were very wicked, brought Jesus to Pontius Pilate, the governor. But Pilate did not want to kill him, because he saw that Jesus was very good and that the Jews were very wicked. 'Crucify him! Crucify him!' the Jews shouted incessantly. 'But he has committed no crime,' said Pilate."[19] This book was published with the approval of the Ministry of Education and the authorization of the Church. In 1940, it was printed in its 58th edition, reaching a total of 50,000 copies.

Another book bearing the same message was *Jesus is the Truth*, which was first published in 1939 with the approval of the Church censorship and the Bishop of Jaen. In January 1941, the book was presented to Pope Pius II with written dedication by the author. In a letter of reply to the author from the Holy See, the Pope praised the book and blessed its author. By 1961, it had already been published in a revised 18th edition.[20]

The same narrative was propagated by books written to help high school students prepare themselves for their matriculation exams. In these books, the "murder of God" is addressed repeatedly, as are other generalizing ideas concerning "the Jews," who demanded the blood of Jesus from Pontius Pilate. Unlike the books designated for elementary school use, these books did not explicitly apply the terms "good" and "wicked" to the characters in the story. Here too, however, the text makes this point abundantly clear: with no other choice, merciful Pilate was forced to do the Jews' bidding.

Historical distortions were not only inserted into books on religion, but also into the 1957 textbooks on the Spanish language. The terminology of a "deicidal people" is present not only in the narratives of Jesus' crucifixion but also in other historical stories unrelated to Jesus.

One such example was a story about Napoleon's treatment of Pope Pius VII. In order to defame Napoleon "who removed the Pope from St. Peter's Throne," and using a concept that was innovative in both content and terminology at the time, the author compared the French sinners during the time of Napoleon to the Jewish sinners during the time of Jesus, who "dragged the Messiah through the streets of Jerusalem" – "the deicidal city" (*ciudad deicida*).[21] The textbooks that were approved by the Ministry of Education on the subjects of art and culture also contained negative information regarding the Jewish people. In these books, every mention of the Jews takes the form of collective condemnation for deicide.[22]

This content is continued in the textbook designated to prepare students for the matriculation exams. According to the curriculum of 1967, the authors also make mention of the sermon and the miracles of St. Peter, who charged the Jews with the abominable sin of crucifying Jesus and threatening his followers with severe punishment.[23] As we have already seen, the major recurring theme in the elementary school textbooks over the years had two primary components: the first was vindication of Pontius Pilate of all guilt with regard to the crucifixion of Jesus, and the second was the placement of all responsibility on the Jews. Although Pilate carried out the sentence, the Romans, who ruled the country at the time, were not primarily responsible for the act: the Jews were. Even when Pilate ordered Jesus to be flogged, his order stemmed from the expectation that the Jews would be satisfied with this punishment, thus enabling him to prevent the murder. The narrative is not identical in all the books. Sometimes it is the Romans who flogged Jesus and placed a wreath of thorns on his head, and at other times it is the Jews. However, in all cases, Pilate is the protagonist trying to save Jesus' life, and the role of the antagonists is played by the Jews. In order to make certain that the young students understood the main point of the narrative, it was summed up concisely in a special frame as follows:

"The Romans carried out the sentence and the execution of God, but the Jews were the inciters and those truly responsible." This message is also reiterated in the pre-formulated question and answer on the subject. The question was "Who was ultimately responsible for the death of Jesus?" The answer was: the Jews.[24]

These ideological trends emerged during the 1950s, a period that witnessed the publication of many books reiterating the link between the Jewish people in its entirety and the enemies of Jesus, including the historical distortions discussed above.[25] For example, one exercise appearing in the well-known textbook *The Yellow Book*, in which students are instructed to complete sentences with the correct answer already provided in parentheses, reads as follows: "He was crucified ______ (by the Jews) on a ______ (cross) made of ______ (wood), between two ______ (thieves)."[26]

Even at the conclusion of the Second Vatican Council in 1965, the Jews, who as we have noted were always referred to collectively, were described using derogatory terms such as cruel, merciless, and instigators of intentional suffering. For example, the Jews "ridiculed" the Messiah and "perpetrated" and caused his "suffering and death." They were also responsible for placing a guard detail around Jesus' grave – they and not the Romans, who are completely absent from the historical narrative presented in the books.[27] In this way, the Jews emerge as a source of threats and intimidation and as seeking to impose terror that must be overcome before the teachings of Jesus can be disseminated.[28]

II. Historical Distortions: "Wicked Pharisees" and Modern Antisemitism

The Spanish textbooks distorted not only historical facts but also the version of the events contained in the Holy Scriptures accepted by the

Christians themselves: the Old Testament and the New Testament. The distortions introduced were intended to better reinforce the standard stereotypes, including those rooted in the assertions of modern Antisemitism that do not appear in the ancient historical Christian sources. What all these distortions have in common is their use of derogatory description and anachronism.

For example, many books describe the Pharisees not as representatives of a stream of Judaism during the period of Jesus but rather as contemporaries of the present and a political party distinguished by "fraudulent behavior." According to the textbooks, the Pharisees claimed to have been "representing and defending the people, but were actually acting to exploit the innocent Jews," and not merely out of hatred for Jesus. One of the recurring arguments against them was that "they hated the Romans," "acted to incite collective hatred of Rome," and were therefore guilty of the Jewish People's hatred of Jesus. In this way, hatred of the Romans was now added to attributes of the Jewish People that were worthy of condemnation. This description of the Pharisees was therefore not only completely detached from the historical reality of the Jewish People living under the burden of Roman occupation, but from the Christian sources themselves. Jewish resistance against Rome was portrayed as unjustified hatred and the Jews' aspiration for liberation was cast as a desire for revenge, through a complete inversion of the role of the Romans, from conquerors to slaves of the Jews. It was Rome that was subject to the influence of the inciting, hateful and vengeful Jews. In the resulting inverse image, the Jews were the powerful hatchers of plots to take over the world, reflecting descriptions taken directly from *The Protocols of the Elders of Zion* and modern Antisemitism. In this way, the textbooks of the Franco era transformed the Jews from the victims of the Roman regime and seekers of independence into a group determined to gain control over the Roman Empire.

The picture that emerged from the textbooks was one of Rome versus Jerusalem, with Rome in the modern world symbolizing Christianity and Jerusalem symbolizing the enemy. The result was a jumbled ideological fusion between the ancient and modern.

These descriptions, which had not even the slightest basis in the Christian Holy Scriptures, were not limited to brief comments but rather filled many pages of the textbooks. They were aimed at imbuing school children of the Franco era with a sense of sympathy for the victorious Empire that accepted Christianity and, at the same time, at propagating hatred of the Jews and casting them as enemies. The nationalist message that emerges depicts a world of good versus evil, with the identity determined by the political-ideological camps that were active in Spain following the Civil War. The struggle between Jerusalem and Rome was symbolic of the opposing camps. For this reason, Jerusalem was described as the "deicidal city" (*Jerusalém deicida*), an overarching symbol of betrayal and evil. The Jewish leadership was charged with responsibility for the murder of Jesus and for his continued torture after crucifixion, as the members of the Sanhedrin who demanded his death "continued their cursing."[29] According to another book, "the Jews hated Jesus to his death out of envy."[30] The Pharisees, then, were envious, and this explained their great hatred of and conspiracy against Jesus. Pilate is referred to only a few times by title –the Roman governor– and even then only in passing, detached from any historical context.[31] The books on Catholic dogma also always refer to the Jews as a collective. The Jews sought the death of Jesus "because they knew that he was about to proclaim himself God."[32] Most of the books in question were published only after being authorized by the official Church censor.

Another subject addressed by the textbooks was the Jewish people's dispersion among the nations, which is explained as part of their divine punishment. The textbooks used by Spanish children during the period give the impression that at the time of Jesus, the Jews ruled their own land and were in control of their own destiny. "The divine punishment that followed the crucifixion of Jesus was realized after the Jews, incited by prophets of fury, rebelled against the Romans and were exiled from their land."[33]

Also prominent in the textbooks were elements that combined the crucifixion of Jesus with concepts that originated during the period of modern Antisemitism and a disregard for historical context. For example, the author of the first grade elementary school textbook *We Saw Our Lord*, chose to use the term "the wicked Jews" (*malvados, malos*).[34]

Another book on Catholic dogma designated for sixth grade elementary school students refers to the Christians as the new divine people, who received the divine blessing that had previously been enjoyed by the Jewish People who had lost it. The Jewish People, it maintains, was "nothing more than a symbol of the new divine People, the Church of Jesus." The text continues as follows: "In his death, Jesus made a covenant with God that was better than the previous covenant, because it was the source of better things for human beings."[35] All this was meant to lead the young reader to clear conclusions regarding the punishment exacted upon the "wicked Pharisees," a term that was synonymous with the Jewish people as a whole.

In one volume of the *Encyclopedia (Escolar Estudio)* Pharisees are referred to as the "wicked Pharisees" (*Los malvados fariseos*). "As an ungrateful People," maintains another book written for young elementary school students, "the Jewish people is deserving of the greatest and most severe punishments."[36] Condemnations of the Jewish People and historical distortions are also widespread in other volumes of the Encyclopedia designated for older children.[37] The Pharisees are always referred to negatively and collectively, and are charged with "maintaining religious ritual ostentatiously and without the inspiration of the true divine spirit."[38] In the religious books for high school students that dealt with the history of the Catholic Church, the Church is characterized as "the true divine People" (*Verus Israel*) and the heirs of the eminence of the Jewish People. Although this description was a common theological explanation throughout the Christian world aimed at displacing the Jews from the divine blessing and birthright, it appears to have received more distinct expression in Spain than in other Spanish speaking areas at the time.[39]

Another dynamic that was common in Spain and devoid of theological grounding was the tendency to refer to the Jews using derogatory epithets. "The Jews" (again, always referred to as a collective) were characterized as being of "bad demeanor," "envious," "hot tempered," and as having "behaved wickedly toward the first Christian sect as a whole," and not only toward Jesus. The first Christians, therefore, were also victims of the Jews, and the Romans, as in the case of Jesus, had no choice but to consent to their demands. For example, in the story of the persecution of St. Esteban, who helped the students of Jesus in different ways, "the Jews, intoxicated with anger, expelled him from the city" and also demanded "the death of Peter." In this way, the Jews were now accused of persecuting not only Jesus but also other Christians whom the Romans sentenced to death.[40] Similarly, in other textbooks used to prepare high school students for the matriculation exams, the Jews were referred to as the murderers of Jesus and the murderers of holy Christians. In this way, they presented and continued to present a great danger that was present and ongoing.[41]

The damning identification of the Jewish People as a whole with the "enemies of Jesus" was advanced in many high school textbooks on the history of the Church, which reiterated all the anti-Jewish stereotypes discussed above.[42] A large portion of the textbooks dealing with religion continued to accuse the Jews of deicide throughout Franco's rule, despite the 1965 change in the Church's official position toward the Jewish people in the *Nostra Aetate* declaration.

A comparison between the elementary school and high school textbooks on religion reveals that the younger the students for whom the books were designated, the more distorted the accounts of the story of the gospels in the Catechism they contained. That is to say, the younger the child the more simplistic, fictitious, lacking in nuance, detached from historical context, and, consequently, powerfully Antisemitic the accounts were that they received. As a result, the children were given the impression that they were dealing with Jewish wickedness not only in the past but in the present, and which posed a threat to them in the here and now. The authors felt the need to pepper their stories with the adjectives "good" and "wicked," thereby presenting a black and white polarized picture in which "the Jews were very wicked" for eternity and Jesus, the symbol of Catholicism, was "very good." Pilate also belonged to the "good" camp, and the absence of any shades of grey in the story limits it to a strictly dichotomous view. "Pilate saw no fault in Jesus" and repeatedly "tried to persuade the wicked Jews" that "Jesus was guilty of no crime." He "wanted very much to release him" and tried "to arouse the mercifulness" of the Jews, but all his good acts proved useless in the face of "the evilness of the Jews." However, in the presence of the cries of "Crucify him! Crucify him!" emanating from the crowd around him, Pilate was left with no other choice but to "order his crucifixion." In this version, the narrative is overly simplistic. Jesus and Pilate are both on the side of the good, and the Jews alone are left on the side of the wicked. Here too, historical context simply melts away. The word "Rome" appears almost nowhere. Children raised on this story would have no way of knowing that Pilate was a governor of the Roman Empire, but knew well that those who ridiculed, struck, and tortured Jesus the entire way to the cross at Golgotha were "the wicked Jews." Those who helped him along the way are not identified as Jews but rather by their first names, as if they were not Jews at all. Jesus himself was also removed from the context of the Jewish People under Roman occupation. The outcome was a story that arouses revulsion and hatred toward the Jews, with the greatest and most overly simplistic distortions and the most glaring anti-Jewish dimensions reserved for the youngest students of all.[43]

It is interesting to note that extreme anti-Semitic pronouncements articulated by Church clergymen were not considered reason to removing Antisemitic priests from positions within the Church censorship. This situation continued until the 1960s. For example, Juan Tusquets, a Catalan priest and Antisemitic author, was also permitted to serve as a textbook censor. Juan Tusquets was not an ordinary priest – he was officially affiliated with the Franco regime, and was appointed as a censor of religious books in Barcelona in 1961 with the authorization of the Archbishop of the city. In 1966, one year after the Second Vatican Council, he approved a book titled *The Bible in Colors*, which was written specifically for young children.[44] Years earlier, in 1938, he had published a series of books attacking the Freemasons and the Communists, as well as a book attacking the Jews, in which he accused them of conspiring against the Catholics with the Freemasons and the Communists. He also charged the Jews with responsibility for the revolution in Spain that resulted in the Spanish Civil War. He frequently quoted the Christian scriptures and began his book with the words of Cardinal Gomá y Tomas, Primate of Spain, who issued a pastoral letter on January 30, 1937 in support of the Nationalist camp's uprising against the Republic. This document, titled "The Christian-Spanish Meaning of the War" (*El sentido Cristiano-español de la guerra*), was also Antisemitic in content, as reflected in the following excerpt, in which the Archbishop blames the Jews for being one of the causes of the 1936 Spanish civil war:

> How can we ignore the responsibility of those who brought here Communism, which is not only an anti-Spanish regime but an anti-human one? What injury has been caused to Spain by those who conspired directly with the Jews and the Freemasons, the true representatives of anti-Spain, who have brought us to this difficult time?[45]

III. Blood Libels

Blood libels –historic accusations linking Jewish ritual customs with the use of Christian blood– are addressed repeatedly in the Catholic textbooks in Spain during the period in question. As we know, a number of Popes rejected claims that Jews used Christian blood for ritual purposes and condemned such stories as being inconsistent with the true nature of the Jewish religion. Nonetheless, blood libels continued to exist in popular Christian tradition and were recounted even after the Pope's pronouncements in the Middle Ages. One example was the blood libel involving a boy named Dominguito de Val of Zaragoza, in 1250, which appeared in one of the textbooks under the title "Blood of the Innocent." According to the version of the story presented by the author:

> At that time, there were many Jews in Spain. Some hated the Christians and were greatly angered by the fact that the children loved Jesus and his mother. For this reason, they killed Dominguito de Val. Dominguito lived in Zaragoza and wanted to be a ship captain. He was a good boy, gave food to the hungry, was brave and defended weak children, and engaged in acts of human kindness. He would walk through the streets of the city singing songs with his friends in honor of the Virgin Mary, the mother of Jesus. He possessed strong

> faith and loved Jesus and his mother, and that is what angered the wicked Jews most of all. As a result, they decided to kill him. One evening, as he passed by the home of one of them, they caught him and took him into his home where they confined him. In the middle of the night, many wicked men entered the house, took the cross that Dominguito carried with him, and ordered him to step on it. Though only seven years old, he offered a brave response: "I will never do it! Jesus is my God!" "Then just as your God died," said the men, "you will die too." They placed a wreath of thorns on his head, nailed him to the cross, and stabbed him through the heart with a spear. Dominguito died on the crucifix, like Jesus, our lord and savior.[46]

Alongside this text, which was written especially for young children who had just learned how to read and write, appeared frightening illustrations. To make it easier for the children to read, the text was printed in large font. It also contained a large picture of Dominguito nailed to the cross, which was consistent with the description of the crucifixion of Jesus. The illustration pictured the young boy on the cross, as four elderly Jews engaged in the act of crucifixion, two hammering the nails into his hands and feet, and the other two collecting the blood that flowed from his wounds. The boy in the picture appears to have been made to look like a typical Spanish child, thereby encouraging the young readers' total sense of identification.

Like other stories of blood libels recounting the crucifixion of Christian children by Jews and the use of their blood for Jewish ritual purposes, the accusations surrounding Dominguito de Val was also denounced by the Papacy in the Middle Ages. Nonetheless, the book was approved for publication in 1940 and was subsequently approved on numerous other occasions, concluding with the approval of a 24th edition in 1961. It was put into use once again in state and private schools by order of the Ministry of Education on March 8, 1961, when it again received Church approval. Its author, Agustín Serrano de Haro, also held the position of Inspector of Elementary Education.

Other textbooks of the period also refer to the Jews collectively as wicked and as the murderers of God and of young Christian children. During the 1960s, Catholic religious elements continued to publish dozens of textbooks that were anti-Jewish in character and that incorporated elements of modern Antisemitism.[47]

In 1967 and 1968, studies were undertaken to locate expressions of anti-Jewish sentiment and modern Antisemitism in Spanish textbooks, yielding two reports on the subject by the Jewish-Christian Friendship Organization. The research revealed twenty-nine titles of a religious nature containing Antisemitic messages that were published during the period in question, some of which were explicitly denounced by the Second Vatican Council in 1965. They were published by the National Institute for the Spanish Book and were all approved by the Church and by the Spanish Ministry of Education and Science.[48]

IV. Textbooks Glorifying the Catholic Monarchs' Expulsion of the Jews in 1492

The negative images of the Jews they contained pertained not only to the period of Jesus but also to modern Spanish history. One issue that became important to many Catholics in Spain during the 1930s and 1940s was the Jewish policy of the Catholic monarchs in 1492, a subject that was addressed by textbooks on both the elementary and high school levels.

Under the circumstances at the time, the expulsion of the Jews from Spain was characterized as a justified, even ingenious, policy toward the

Jews. According to the narratives advanced by the textbooks, the policy implemented by the Catholic monarchs, King Ferdinand and Queen Isabella, was justified, despite the criticism to which it was subject at the time. This assertion, however, was meant only to serve as background for the primary message being conveyed to the subjects of Francoist Spain: equivalent justification for the persecution and expulsion of Jews from Spain and the persecution of Jews in Nazi Germany. According to the textbooks, the expulsion was a preventative measure, a means of protecting Spain from the Jewish threat, and for this reason the Christian policy was warranted. As Christians are forbidden to commit murder, expelling the Jews from Spain was a legitimate alternative, and was portrayed as "patriotic and just" from a Christian perspective as well. According to the textbooks, the expulsion was not violent and did no physical injury to the Jews. Neither was it accompanied by acts of murder nor was it intended strictly as an act of self-defense. Whereas violence and murder were considered as running contrary to Christian doctrine, the policy of expulsion as a defensive measure was an integral component of the message conveyed to children by the textbooks as an example of a perfect Christian solution.

The solution of expelling the Jews employed by the Catholic monarchs, Ferdinand and Isabella, in 1492 also served as a basis on which to draw conclusions regarding the period of Franco. Jews should not be allowed to return to Spain because they were dangerous. Moreover, Spain was not tainted by racism and there was no reason to import any, especially when countries the world over were attempting to rid themselves of Jews, who were welcome nowhere.[49] "The Catholic monarchs expelled the Jews, who were a true national threat," read one elementary school textbook. "The institution of the holy Inquisition was established to protect the unity of the faith."[50]

The books described the Catholic monarchs, who unified the kingdoms of Aragon and Castile in 1492, as leaders who achieved national unity and halted the Muslim *Reconquista* of Spain with the victory over the Muslims (Moros) in Granada in 1492. They also credited Ferdinand and Isabella with saving Spain from the Jews. Although the Jews were described as having lived in Spain before that time, the books emphasized that they were hated by the Spanish and that their expulsion reflected the great wisdom of the monarchs and proved religiously, nationally, politically, and economically beneficial to the country.

> In neighborhoods known as *Juderías*, they [the Jews] engaged in usury. The people hated them and charged them with deeds of the most terrible and abhorrent nature… In order to achieve religious unity, they [the Catholic monarchs] brought about the Jews' removal from Spain because they were hated by the people. The expulsion of the Jews benefited Spain both religiously and politically.[51]

The story of the Spanish expulsion, then, was recounted in the context of Spain's aspiration for religious unification and the general hatred of the Jews stemming from their economic transgressions. In addition to fulfilling this aspiration, the Catholic monarchs were credited with another important achievement: the establishment of the Inquisition, which was aimed at "identifying heretics and Jews who were pretending to be Catholics."[52] The expulsion of the Jews was portrayed as part of this effort, and, as already noted, provided a solution for the true feelings of the people.

According to the authors, the golden age of Spain began with the marriage of the Catholic monarchs of Aragon and Castile and the unification of Spain in the fifteenth century. Ferdinand and Isabella overcame the anarchy that had prevailed up to that point, for which the

Jews, among others, had been responsible. In the words of one textbook: "They expelled the Jews, the true worms of the nation."[53] Many of the books offered an interpretation of the expulsion aimed primarily at proving that the Jews were the enemy of Jesus.[54] The expulsion from Spain was also linked to the establishment of the Inquisition, which was initiated to fulfill the vision of Catholicism as the only religion of the united country. To this end, the kingdom expelled the Jews and undertook the Inquisition against the heretics. "These measures," another book read, "were what made Spain the [world's] most important country during that period."[55]

V. Defending the Morality of the Inquisition

The total conception that presumed the existence of a clear distinction between the "good" and the "wicked" was not applied to the Jews alone. It was manifested in the treatment of other religions, such as Protestants, Orthodox Christians, and Muslims. It was also applied to political issues, perhaps the most vivid example of which was the treatment of the Spanish Civil War. In this context, Franco was cast as a savior rescuing Spain from its internal enemies – the communists, or the "reds," as they were referred to. The causes of the Civil War were not attributed to the Jews alone, but also to the Freemasons, the socialists, and the communists. The textbooks referred to Franco's uprising against the elected democratic republic as the War of Liberation (*Guerra de Liberación*), and in this context, too, clearly distinguished between the "good" and the "wicked." For example, they explained how, in honor of Franco's steadfastness against the enemies on the home front –the communists, the Freemasons, and the Jews– José Antonio Primo de Rivera, the founder of the Spanish fascist Falange party, earned the status of "national hero," along with General Francisco Franco, for delivering the country from them.[56]

The textbooks dealing with the history of the Catholic Church depict the Inquisition as an institution set up to fight heresy. The textbooks stressed that discussion of heresy began in the Middle Ages, when it was considered not merely a religious crime but a social crime as well. According to the textbooks, the Inquisition was established during the thirteenth century at the request of the Popes from southern France to investigate who was responsible for disseminating deviations from the faith. As an official body, the Inquisition operated in France, Spain, Italy, and Germany. But in Spain it operated longer than elsewhere and was more effective. As an institution, it was credited with saving Spain from heresy and religious wars that reaped death and destruction, like those that had plagued other countries in Europe. Because of the Inquisition, Spain enjoyed many advantages, including political and religious unity.[57] "In this way," one book explains, "it freed our country from many of the evils of the modern world."[58]

The textbooks described the Inquisition in the most sympathetic terms and vindicated it of all wrongdoing, characterizing it as a supervisory arm responsible for religious and civil duties. According to this account, and attempting to present an apologetic view of the Church, the textbooks stressed that the punishments it assigned were authorized by the civil leadership and not by the Church itself. The textbooks expressed total admiration for the Church's actions and official decisions, and were devoid of any criticism or doubts on the subject.

Conclusion

In the realm of foreign policy, part of the historiography of WWII depicts Francoist Spain as a force responsible for saving Jews during the war. But historians are divided on this issue. There is consensus regarding the fact that the Jews who fled the Nazis and made it to Spain were indeed saved. The disagreement has to do with the scope, motivation, and significance of the policy of rescue, whether or not Spain exhausted its potential to rescue Jews, and if not, why?[59] The Franco regime was economically dependent on the West and was preparing itself for the post-war era. To this end, it sought ways in which to improve its relations with the Allied Powers. Providing assistance to Jewish refugees was one way of achieving this goal. At the same time, 450 years after the expulsion of Jews from Spain, the country was still cautious about the possibility of mass Jewish immigration into its territory and took action to prevent it.[60]

Numerous statements issued by the Spanish Foreign Ministry during and following WWII emphasized that Spanish law did not permit discrimination on the basis of race or religion.[61] However, even if no racially Antisemitic laws were ever enacted in Spain as they were in other European countries, the Spanish government did issue orders aimed at limiting the presence of Jews and keeping their numbers to a minimum.[62] Laws enacted during the Franco era permitted the sweeping persecution of Freemasons and communists, who were classified as ideological and political enemies.[63] These laws also provided the basis for regime efforts to limit the Jewish presence in the country. Regime officials justified this limitation based on their classification of the Jews as dangerous religious and political enemies.[64] On the other hand, some Spanish diplomats at the time appealed to Spanish Catholic sentiment in an effort to save Jews. One example was the Spanish consul in Berlin, who appealed to the Spanish Foreign Ministry to act to save Jews "in the name of Catholic compassion and justice" and to do its utmost to assist in the absorption of Jewish refugees in Spain in order to save them from certain death.[65]

Spain's education system was not reflective of the country's changing foreign policy regarding the Jews. An assessment of the policies of the education system as reflected in the textbooks of the period indicates a system that operated independently of the Foreign Ministry and that continued implementing the regime's intentionally repressive policies toward the Republican camp defeated during the Civil War. Jewish stereotypes constituted an integral element of this dualistic ideological indoctrination, which highlights the ideological contradictions between the two camps and the policy of obliterating all vestiges of the Republican era and its values. Such content permeates the textbooks and their treatment of the Jews in particular. During the period in question, when the Catholic Church was responsible for educational and academic content, certainly in the realm of religion, textbooks were replete with views that were injurious to the Jews.

During the era of Franco's authoritarian regime in Spain that followed the Republican victory in April 1939, the textbooks which the state education system provided to its students reflected the Catholic Church's position vis-à-vis the Jews. In the realm of education, the Catholic Church enjoyed supremacy over the other ideological and political groups that were party to the regime. The Falangists and the integrist Catholic Monarchists may have held important ministries, but the Catholic Church exercised complete control over the Ministry of Education.

By virtue of the Church's control over the education system, textbooks were placed at the service of the ideological and political needs of the regime. This was true of elementary school textbooks, textbooks for the higher grades, and textbooks designated to prepare students for the matriculation exams. The education system purged its textbooks of all vestiges of democratic, liberal, and socialist values. At the same time, it cultivated values of Catholic religious tradition and Spanish Imperial tradition and combined them with content advanced by Spanish fascism –by the Falange party and its founder José Antonio Primo de Rivera, and, first and foremost, by General Francisco Franco, the leader of the regime, who was depicted as having rescued Spain from its internal and external enemies. This content was incorporated both into the textbooks used within the general state education system and the private Catholic education system. It appeared in textbooks on various subjects, such as history, geography, literature and the Spanish language, as well as textbooks dealing with religion.

The regime's attitude toward the Jews during the period must also be understood against the backdrop of Spain's ideological education system and its role as a tool for the indoctrination of the young minds of the country. The findings of this study indicate that during the years in question, Spanish textbooks were dominated by Anti-Jewish Catholic conceptions that emphasized the Jews' responsibility for murdering Jesus. This approach was not unique to Spain and was not unusual in the Catholic world of the 1930s and the 1940s. Indeed, it was part of a way of thinking that was widespread in different variations throughout the Catholic world and that received different emphasis in different times and places.

The unique nature of Antisemitism in Francoist Spain that was manifested in the textbooks discussed in this article was rooted not only in former anti-Jewish Christian religious tradition but also in the emphasis on unified Spain's glorious past and imperial and Catholic values, which included the expulsion of the Jews and support of the values of the Inquisition. The innovation came in its scope and terminology. Previous limits of religious subjects were expanded, as were the antisemitic stereotypes employed. Alongside familiar stereotypes from ancient times and the Middle Ages, authors now added new concepts originating in Spanish history and in the collection of images provided by modern Antisemitism.

As we have seen, these later editions included elementary and high school textbooks, as well as books designated for preparing high school students for the matriculation exams. Spain was a country that made religious education a mandatory subject of study on all levels and in all schools, state and private alike, which meant that the children of Spain were educated in Antisemitic stereotypes from birth until adulthood. The textbooks also endowed Spanish schoolchildren with tools to justify the deeds perpetrated during Spain's Imperial and Catholic past, from the era of the Catholic monarchs onward.

The signing of the Concordat between the Vatican and the Spanish government in 1953 strengthened the Franco regime by endowing it with the legitimacy of an official agreement with the Pope and thereby indirectly bolstered the legitimacy of its anti-Jewish views. According to the Concordat, the Roman Apostolic Catholic Church would continue to enjoy its special rights under divine law and canonical law, an arrangement that remained in place until the change in the Constitution in 1978.[66] Textbooks with Antisemitic content continued to be published throughout the 1950s and 1960s, in spite of that after WWII the horrors of the Holocaust were publicly known to all in Spain.

Furthermore, in 1965 the *Nostra Aetate* declaration composed by the Second Vatican Council, did away with the collective accusation against the Jewish people for deicide against both the Jews from the days of Jesus and those that lived after him.[67] Nonetheless, according to the books surveyed for this study, the negative stereotypes of the Jews continued to appear in the textbooks of Francoist Spain until at least 1969, until the results of research conducted under the leadership of the nuns of the Congregation of Zion revealed the Antisemitic educational content in the textbooks that were in opposition with the new *Nostra Aetate declaration.*

As we have seen, the change in the Vatican's position regarding the Jews made at the end of the Second Vatican Council in 1965 had no impact on the education system of the Franco regime, which continued to disseminate and reprint the existing textbooks, and the Antisemitic material they contained, in complete disregard of the substantive change which the Church had agreed to take upon itself throughout the world. The textbooks continued to nurture all the negative stereotypes that had become a fixed feature of traditional Catholic anti-Jewish literature, as well as those taken from modern Antisemtism. These textbooks –which were written for all the children of Spain in the spirit of the Catholic Church, whose representatives controlled the Education Ministry during the entirety of Franco's time in power– were permeated by a distinct Antisemitic theme. According to the message that emerged, Spain's treatment of the Jews was a distinct instance of a country taking preventative action to head off the harmful influence of Judaism and the Jews. From this perspective, the expulsion of the Jews from Spain by the Catholic monarchs Ferdinand and Isabella in 1492 was seen an example of wise, exemplary behavior. During the war, immediately following the war, and in the subsequent two decades, the books made no mention of the Holocaust. During this period too, Antisemitism continued to serve as an integral part of Spanish education under Franco and the textbooks authorized by the state Education Ministry and the Church censorship.

As already mentioned, a reassessment of Spanish textbooks was carried out as a result of pressure exerted by nuns of the Sisters of Zion, who, following the Second Vatican Council, called for presenting the issue to a commission of the Ministry of Education to prompt it to order the revision textbooks in accordance with the new instructions contained in the *Nostra Aetate*.

Neither reports regarding the Nazis' persecution of the Jews in Germany during the 1930s, nor news about the mass killing of Jews under German Nazi rule in Europe during the war affected the Antisemitic narrative in the Spanish educational system and its textbooks, which continue to use the same kind of rhetoric after the Second Vatican Council as it did at the beginning of the Second World War.

Notes

1 Haim Avni, *España, Franco y los judíos* (1st ed. in Hebrew, 1974; Madrid, 1982); Antonio Marquina y Gloria Inés Ospina, *España y los judíos en el siglo XX* (Madrid, 1987); José Antonio Lisbona, *Retorno a Sefarad. La política de España hacia sus judíos en el siglo XX* (Barcelona, 1992); Raanan Rein, *In the Shadow of the Holocaust and the Inquisition* (Tel Aviv, 1995), 76-77 (Hebrew). Pascal Blin "Franco et les Juifs. Paroles et actes," Ph.D. Dissertation, Universite de la Sorbonne Nouvelle-Paris III.

2 William Callahan, *La Iglesia católica en España (1875 – 2002)* (Barcelona, 2003), 303-320; José Díaz Gijon, Donato Fernández Navarrete et al., *Historia de la España actual 1939 – 1996. Autoritarismo y democracia* (Madrid and Barcelona, 1998), 57–78.

3 Julián Casanova, *La Iglesia de Franco (*Madrid, 2001), 274.

4 See Graciela Ben-Dror, "Trends within the Catholic Right in Spain: Viewing the Other as an 'Enemy of Spain' and as a Traitor to the Homeland (1936-1945)," in Meir Litvak and Ora Limor (eds.), *Religious Fanaticism* (Jerusalem, 2007), 329-357 (Hebrew).

5 Danielle Rozenberg, "Minorías religiosas y construción democrática en España (Del monopolio de la Iglesia a la gestión del pluralismo)," *Revista Española de Investigaciones Sociológicas* 74 (April-June 1996), 248-249.

6 See: Graciela Ben-Dror, *La Iglesia Católica durante el Holocausto: España y América Latina* (Madrid, 2003).

7 About *Nostra Aetate* see: "Declaración Nostra Aetate sobre las realciones de la Iglesia con las religiones no cristianas", *Vaticano II, Documentos Conciliares* (1st ed. 1981, Buenos Aires: Ediciones Paulinas, 1984) 465-471.

8 Carolyn Boyd, *Historia Patria* (Princeton, 1997), 261-262.

9 Ibid., 257-258.

10 Ibid., 262.

11 José María Pemán, *Manual de Historia de España* (Santander, 1939).

12 Luiz Ortiz Muñoz, *Glorias Imperiales: Libro escolar de lecturas históricas*, 2 vol. (Madrid, 1940).

13 See: Agustín Serrano de Haro, *Yo soy español* (Madrid, 1946), 6.

14 On the changes in the Catholic Church, see: Frances Lannon, *Privilege, Persecution and Prophecy, the Catholic Church in Spain* 1875-1975 (New York, 1987). On liberalization, see : Javier Tussell, *La oposición democrática al franquismo, 1939-1962* (Barcelona, 1977); Elías Díaz, *Pensamiento español en la era de Franco (1936-1975)* (Madrid, 1983).

15 Eldevives, *Historia Sagrada, 2do grado* (Zaragoza, 1961). 194-196, Imprimatur; also: Eldevives, *Jesucristo según los Evangelios* (Zaragoza, 1966), 146-147, 121.

16 Federico Bassó Ubach, Pbro, *Creo en Jesucristo, Mi cuaderno de Catequesis* (Barcelona, 1962), 71. This book was printed with the authorization of Church censor Dr. Quirico Estop, Pbro. Barcelona, July 19, 1962 and Archbishop Gregorio of Barcelona. A fourth printing of the book was released in 1966.

17 José Garmendia Galdós, S.M., *Historia Sagrada, Curso elemental, Apropiado para 2 o 3er grado* (Madrid, 1965), 98-99.

18 Juan Navarro Higuera, Inspector de Enseñanza Primaria, *Lecciones Amenas, Segundo libro, para niños de 7 a 9 años* (Barcelona, 1959), 54. This book was approved by the Ministry of Education and the Catholic Church.

19 Agustín Serrano de Haro, *Hemos visto al Señor, El libro de primer Grado de Religión* (Madrid, 1940), 58-61. See also *Nueva Enciclopedia Escolar, Grado Tercero de Enseñanza Primaria* (Burgos, 1965), which received Church approval from censor Marcelino Martínez Hernado and Archbishop Luciano of Burgos (May 18, 1960).

20 Agustín Serrano de Haro, *Cristo es la Verdad* (Madrid, 1961), 99.

21 *Lengua Española, Segundo curso de Bachillerato, Plan 1957* (Madrid, 1957), 103. This book was published with Ministry of Education approval on September 30, 1963.

22 Demetrio Ramos, *Estilo, Historia del arte y de la cultura* (Barcelona, 1960), 61.

23 Mariano Villapún Sancha, *Historia de la Iglesia., Tercer Curso de Bachillerato* (Madrid, 1967), 8-9. Plan 1957.

24 Federico Bassó Ubach, Pbro., *Creo en Jesucristo. Mi cuaderno de Catequesis* (Barcelona, 1965), 71-73.

25 *Perfeccionamiento, Libro IV de labor escolar* (Gerona, 1957).

26 *Enciclopedia Escolar Estudio* (Libro amarillo) (Gerona, 1964), 37.

27 *Nueva Enciclopedia Escolar HSR Grado Segundo, nueve-diez años* (Burgos, 1955), 68-69. This reference appears in the updated 53rd edition.

28 Eutiquio Peña Rica, Manuel Useros Carretero (dir.) *La doctrina de Jesucristo, Cuarto Curso* (Madrid, 1965), 56.

29 J. Mallas Casas, *Humanismo y Progreso* (Barcelona, 1963), 72-80.

30 Antonio Fernández Rodríguez, *Enciclopedia práctica, Grado Medio* (13th ed., Barcelona), 89. This book emphasizes that it was published with the required Church approval.

31 Ibid.

32 Eutiquio Peña Rica, Manuel Usero Carretero. *El Dogma Católico, Sexto Curso* (Madrid, 1967), 79.

33 Eldevives, *Historia Sagrada, Segundo Grado* (Zaragoza, 1961), 195, 196, 197, 198, 199, 224-225.

34 Agustín Serrano de Haro, *Hemos visto al Señor, El libro de primer grado de religión* (1st ed. 1940, 58th ed., Madrid, 1961).

35 Eutiquio Peña Rica, Manuel Usero Carretero, *El Dogma Católico, Sexto Curso* (Madrid, 1967), 118.

36 *Enciclopedia Escolar Estudio, Libro Verde* (Gerona, 1962), 74-80; 82-84.

37 *Enciclopedia Cíclico-Pedagógica, Grado Medio* (Gerona, 1962); *Enciclopedia. Grado Superior* (Gerona 1963); *Enciclopedia escolar Estudio. Libro Azul* (Gerona, 1958); *Enciclopedia escolar Estudio Libro Rosa* (Gerona, 1962), 128, 122, 120, 118, 117.

38 Mariano Villapún Sancha, *Jesucristo según los evangelios, Segundo Curso Bachillerato, Plan 1957* (Madrid, 1967), 24 Eutiquio Peña Rica y Manuel Useros Carretero, *Jesucristo según los Evangelios, Historia Sagrada del Nuevo Testamento*, Segundo Curso (Madrid, 1965), 130- 131, 146-150. Nihil Obstat , Pablo de Olmo, May 29, 1965 and Imprimatur of Segundo, Arzobispo de Burgos. See also: Alvarez y C. Herrero, *Religión, Sexto Curso* (Valladolid, 1967), 64-75 (About the Crucifixion) Nihil Obstat Lic. Martin Gil Diez, Valladolid, June 8, 1967, Imprimatur José, Arzobispo de Valladolid. 1st. ed. 1967.

39 Eutiquio Peña Rica, Manuel Useros Carretero (Dir), *El Dogma Católico, Sexto Curso* (Madrid, 1967), 119. Text adapted to the questions of the official de education, approved July 17, 1965. With the Licencia Eclesiástica . Nihil Obstat of Dr. Lamberto de Echeverría, Censor, and Imprimatour of Dr. Constantino Palomo Vicario General , Salamanca, May 25, 1965. See also for example a comparative study of the Catholic Church in some Spanish speaking countries in Graciela Ben-Dror, La *Iglesia Católica y el Holocausto: España y América Latina* (Madrid, 2003).

40 Eutiquio Peña Rica, Manuel Useros Carretero (dir), *Historia de la Iglesia. Tercer curso* (Madrid, 1967), 28, 33, Nihil Obstat of Lic. Pablo del Olmo, Burgos, May 7, 1965 and Imprimatur of Segundo, Arzobispo de Burgos.

41 *Historia de la Iglesia y Liturgia, Tercer año de Bachillerato* (Madrid, 1967), 14 Plan 1957. Text aproved by the Ministerio de Educación Nacional, O. M., January 10, 1964, 14.

42 Mariano Villapún Sancha, *Jesucristo según los Evangelios. 2do. Curso* (Madrid, 1958); Idem., Villapún Sancha, *Historia de la Iglesia. 3er. Curso* (Madrid, 1959); Villapún Sancha, Idem., *La Doctrina de Jesucristo. 4to Curso* (Madrid, 1960).

43 Agustín Serrano de Haro, *Hemos visto al Señor, El libro de primer Grado de Religión* (1st.ed. 1940, 58 ed., Madrid, 1961).

44 *Biblia en Colores*, Spanish version of José Luis Albizu, about an original german edition of Bilderbibel, about texts prepared by W. Hillman, OFM, with the collaboration of Josef Quadflieg and published by Patmos Verlag de Dusseldorf, 1958. First edition in Spanish with the Nihil Obstat of the censor Dr. Juan Tusquets, Pbro. Prelado Doméstico de S.S. Imprímase: Barcelona, 31 de Octubre de 1961, Gregorio, Arzobipo-Obispo de Barcelona, 1961 (First edition Barcelona 1963, Second edition in 1966 (Barcelona, 1966), 76-77.

45 About Juan Tusquets see: Barón de Santa Clara, *El Judaísmo*, vol. VI, Director J. Tusquets (Burgos, 1938), 5 (With the eclesiastical licence).

46 Agustín Serrano de Haro, *Yo soy español* (1st ed. 1940; Madrid, 1961), 54-56, Nihil Obstat, Lic. Balbino Carrilo, Censor, Jaen 10 Abril de 1943, Imprimatur, Rafael, Episcopus Cienses.

47 Vicente Hernando y Víctor Fernández, *Lecturas Históricas* (Madrid, 1964). Antonio Fernández Rodríguez, *Enciclopedia Didáctica. Período elemental* (Barcelona, 1957), 61, 63; Antonio Fernández Rodríguez, *Estampas de la Stma. Virgen* (Barcelona, 1959), 62.

48 Amistad Judeo-Cristiana. *Informe sobre revisión de libros de texto de enseñanza primaria,* Madrid, June 30, 1967*; Amistad Judeo-Cristiana. Segundo informe sobre revisión de libros de texto*, Madrid, October 31, 1968.

49 Ibid., see also: Carlos Benarroch, "La Real Academia Española y los judíos, *Boletin Amistad Judeo-Cristiana*, no. 31 (June-July-August 1967): 8-11; Vicente Serrano, "El judaísmo en los libros de enseñanza religiosa en España", *El Olivo*, no. 7-8 (December 1978): 7-28.

50 Antonio Fernández Rodríguez , *Enciclopedia Práctica, Grado elemental* (Barcelona, 1964) (Written that it received the Eclesiastical License and approved by the Ministry of National Education).

51 *Nueva Enciclopedia Escolar* H.S.R. *Correspondiente al período de perfeccionamiento diez-doce años* (ed. no. 45, Burgos, 1965), 566-569.

52 Ibid.

53 Antonio Fernández Rodríguez, *Enciclopedia práctica, Grado Medio* (Barcelona, 1964), 29.

54 Ediciones S. M. *Historia Universal y de España. 4to. Curso* (Madrid, 1965); M. Sáenz de Uruturi y V. Mateo Velasco, *Historia de la Literatura 6to Curso* (Madrid, 1965); Juan Vesga y Félix Fernández, *Jesucristo según los Evangelios* (Madrid, 1964), Eduardo Benlloch y César Tejedor, *Camino, verdad y vida. Moral Católica. 5to Curso* (Madrid, 1965), 186, 187, 188). Buenaventura Barrón, *La Santa Iglesia, Su Historia. 3er. Curso* (Madrid, 1967), 9, 102 –105, 129.

55 José L. Otaño, *Geografía e Historia de España. Curso de iniciación, 3er grado, 8 a 9 años* (Madrid,1964), 90-91.

56 Ibid., José García López, *Historia de la Literatura, Sexto curso* (Barcelona, 1962), 77, 94 – 95, Enciclopedia Escolar Estudio.

57 *Historia de la Iglesia y Liturgia. Tercer año de Bachillerato* (Madrid, 1967), 93 Plan 1957. Text aprovedd by the Ministery of National Education January 10, 1964, 13, 14, 21, 30, 45, 77, 83, 94, 104-105, See also: Pla Dalmau, *Enciclopedia Cíclico-Pedagógica, Grado Superior* (Madrid, Gerona, 1963), pp. 468, 469- 470.

58 Eldevives, *Historia de la Iglesia, Religión, Tercer Curso*, (Zaragoza, 1964), 91, With the Nihil Obstat Ferdinandus Pérez Aysa and Imprimatur of Dr. Aloysius Borraz, Vicario Capitular, 1964.

59 Haim Avni, *España, Franco y los Judíos* (1st ed. In Hebrew, 1974; Madrid, 1982); Antonio Marquina y Gloria Inés Ospina, *España y los judíos en el siglo XX* (Madrid, 1987); José Antonio Lisbona, *Retorno a Sefarad. La política de España hacia sus judíos en el siglo XX* (Barcelona, 1992), Raanan Rein, *In the Shadow of the Holocaust and the Inquisition* (in Hebrew) (Tel Aviv, University Publishing Projects, 1995), 76-77.

60 Pascal Blin "Franco et les Juifs. Paroles et actes", Ph.D. diss. Universite de la Sorbonne Nouvelle-Paris III, 1992.

61 See in: AMAE, Leg R 1460, Exp. 14, from 5.12.1938 says there is no racial reglamentation that can ban to extend passports to Sephardic Jews; Orden Circular no. 114, 16.8.1939, quoted in Leg. R. 2153, Exp. 41, from March 18, 1942 remained that there is no legislation against Jews.

62 Ibid.

63 Ma. Carmen García-Nieto y Javier Ma. Donezar, *Bases documentales de la España contemporánea. La España de Franco 1939-1973* (Madrid, 1974), 111-115.

64 This was reinforced by some diplomats, like the Spanish Ambassador José de Lequerica in Vichy, who spoke about the "Jewish menace" *(El peligro judío)*, and said that in spite of the fact there were few Jews entering illegally through Spain's borders, they were very dangerous, because they undo the good work done by the Catholic Kings in 1492, see: AMAE Leg. R 1716 Exp. 4, doc. no 481, Vichy 29.9.1942.

65 A letter signed as Federico, the consul in Berlin, sent to Jose, meaning José Doussinague, General Director of Exterior Policy in the Ministry of Exterior, AMAE, Leg. 1716, Exp. 5, Berlin, 22.7.1943.

66 Ma. Carmen Garcia-Nieto y Javier Ma. Donezar, *Bases documentales de la España contemporánea. La España de Franco 1939-1973*, 20.

67 About *Nostra Aetate* see: "Declaración Nostra Aetate sobre las realciones de la Iglesia con las religiones no cristianas", *Vaticano II, Documentos Conciliares* (Buenos Aires, 1984) 465-471; Regarding some important documents delivered from 1965 onwards by the Vatican II toward the Jews, see: *Cristianos y judíos por los caminos del diálogo*, (Madrid: Comisión Episcopal de Relaciones Interconfesionales, 1992), 11-39.

Boris Morozov & Marina Niznik

The Jewish Community and Electoral Campaigns in Russia, 2007-2008

The main aim of the research we conducted in the Stephen Roth Institute for the Study of Contemporary Antisemitism and Racism, based on material gathered from the Russian press and the internet, was to discover the attitude of the Jewish community and the position of the main Jewish organizations regarding the elections to the State Duma and the Presidential elections that took place at the end of 2007 and the beginning of 2008. We also wanted to analyze to what extent the so-called "Jewish Question" was relevant and used by political parties of various orientations, and to see if there were any pro-Jewish or anti-Semitic statements, slogans or other forms of propaganda during the election campaigns.

Methodology

To analyze the electoral campaigns for the elections to the Russian State Duma (2nd December 2007), as well as the Presidential campaign (2nd March 2008), we examined the Russian press (including the newspapers with state circulation, certain regional newspapers, newspapers of Jewish communities as well as newspapers known for their extreme anti-Semitic views –(see our research on anti-Semitism in the Russian press) during the two months prior to the elections. Though the official period of the electoral campaigns in Russia for both elections lasted only one month, we decided to see if there had been any pro- and/or anti-Jewish statements during the pre-electoral period. At the same time we wanted to check if there was any post-electoral reaction, especially in the various extremist newspapers. In order to see if there was a difference in attitude towards the electoral campaigns in areas with strong Communist representation[1] we took *Ul'ianovskaia gubernia* as an example, and compared its regional newspapers, available in the State Library of the Russian Federation, with the newspapers of Moscow, where "*Edinaia Rossiia*"[2] is clearly dominant. In attempting to discover the attitude of Jews and Jewish communities of Russia towards the elections we researched the newspapers of Jewish communities during the same period. We also attempted to examine SOVA's[3] statements in the same months in order to see if there was any correlation between anti-Semitic activities and the elections. Finally, we also managed to receive data on the results of voting of Russian Jews permanently living in Israel who participated in the elections for the State Duma and Presidential election, organized in Israel by the Russian Embassy. The material we collected is presented below.

The attitude of the Russian Jews towards elections

Researching this question proved somewhat difficult because of the lack of directly corresponding data. Unfortunately, there is no reliable data even regarding the number of Jews living in Russia today. According to the statistics of FEOR (Federation of Jewish Communities of Russia) there are about 2 million Jews in the country, while according to the last census there are only about 230,000.[4]

Of course this brings up the complex question of whom we consider to be a Jew – those who have the right of return to Israel, or those who identified themselves as Jews in the census. In any event, today Jews in

Russia constitute a rather small minority of the total population of the country, and their influence in the general electoral process is practically negligible. Unfortunately, there is no existing data on the results of voting regarding different nationalities or religions. This gives us certain grounds to assume that in general the voting of Jews did not drastically differ from that of other citizens of Moscow and other large cities of the Russian Federation.[5]

The results of censuses conducted during previous elections show that in general there is a tendency for Russian Jews to vote for parties and candidates associated with "democratic forces". For example, according to the data of the Levada Center during the Presidential elections of 1997, the Jewish community supported Yeltsin against Ziuganov;[6] in 2000 and 2004 the vast majority of them voted for Putin [mainstream voting]. As for elections to the State Duma in 1999, the Jews supported liberal and democratic parties against the Communists and national-patriots, while during the previous elections of 2003, the vast majority of the Russian Jewish community voted for the *"Edinaia Rossia"*, "*Iabloko*" and "*Soiuz pravyh sil*" parties against the Communist and national-patriotic parties of the *LDPR* and *Rodina*.

As for the reaction of the leaders of the Jewish community in Russia regarding the results of various elections, it is necessary to mention one of the two Chief Rabbis of Russia, Berl Lazar, who 3 days before the Presidential elections of 2004 announced that every Jewish voter has an obligation to vote, and was satisfied with the results. He said that the elections proved that Russian Jews feel safe. M.Chlenov [General Secretary of the Euro-Asian Jewish Congress] stated that though Russian Jews sympathize with the liberalism of the Western variety, they are ready to integrate into the "vertical power" built by Putin because they are afraid of Russian nationalism as a state policy.

The situation in Russia today, where there is no state anti-Semitism, is naturally preferable for the Jewish communities and they would definitely like to preserve the status quo. This is why it is quite understandable that Berl Lazar said that for the Jews of Russia participation in democratic elections is obeying God's commandment, since according to Jewish tradition elections are the part of God's gift of free choice. He also said that the result of elections is very important for Jews because the achievement of international peace depends on it.[7] As a result of the elections to the State Duma only 4 parties out of 11 managed to overcome the 5% electoral barrier: *Edinaia Rossiia* (United Russia) – 64.30%, Communist Party of Russian Federation – 11.57%, Liberal Democratic Party of Russia (Party of Zhirinovskii) – 8.14% and *Spravedlivaia Rossiia* (Just Russia) – 7.74%). The parties representing the "right" or so-called "democratic" political stream, "*Iabloko*" (Apple) and SPS (The Council of Right Forces) lost the vote.

Pro- or anti-Jewish statements during the electoral campaigns in the mainstream newspapers

The results of both electoral campaigns were quite predictable. Consequently, there were no significant political battles either between political parties during the elections for the State Duma or between candidates for the Presidency. *Edinaia Rossiia,* which had the majority in the previous Duma, was originally supposed to win the elections using the "administrative resource" – meaning the possibilities of ruling structures possessing the power. The question was whether it would be able to reach the level of the so-called constitutional majority (more than 75%), and hence gain the right to overcome the presidential veto.

The elections of 2007-2008 were conducted under conditions where the authorities made it very clear that they would not tolerate any open manifestations of anti-Semitism or xenophobia.[8] This is why, unlike the previous elections, during which anti-Semitic and xenophobic propaganda was used,[9] these last campaigns did not show any significant examples of it.

Only 3 other parties were able to overcome the electoral barrier, and this is why only the Communists, the LDPR and *Spravedlivaia Rossiia* were really active in their propaganda. Out of these three, *Spravedlivaia Rossiia*, which was also considered to be the Kremlin's project aimed to distract the attention of certain groups of potential Communist voters, never seriously struggled with ER and therefore did not manifest any xenophobic tendencies.

LDPR, on the other hand, attacked everybody, from "democrats" to "communists", evidenced by the blunt rhetoric of its leader, Vladimir Zhirinovskii . His accusations against the existing power were very populist; he applied to the common people, presenting LDPR as their only real defender. Comparing the latest campaign of LDPR with the previous ones, we can assume that in general Zhirinovskii tried to demonstrate politically correct behavior. Nevertheless, in his propaganda before the elections to the State Duma he could not refrain from making definite xenophobic statements directed against migrants from Caucasus and Middle Asia.[10] Though this time the leader of LDPR was very careful and tried to avoid direct anti-Semitic statements, one can still detect hidden nuances. For example, in his propagandistic piece entitled "*Russkaia vlast*" (Russian power) he wrote that there is no natural resource in Russia which would not be seized by "*bogoizbrannye*" (selected by God) oligarchs.[11] Incidentally, the slogan "*My – za russkikh, my – za bednykh*" (We are for Russians, we are for the poor), was again used many times in LDPR propaganda, and as usual was accompanied by the statement "*Khorosho russkim – khorosho vsem*" (Good for Russians - good for everybody). We did not find any anti-Semitic manifestations in Zhirinovskii's political proclamations in preparation for the presidential campaign.

The Communists seriously criticized *Edinaia Rossiia*, Putin and the Russian authorities in their major newspapers "*Pravda*" and "*Sovetskaia Rossiia*", as well as in their regional and local editions. But they were extremely careful and politically correct, and never gave the electoral committee the opportunity to accuse them of xenophobia and thus stop their campaign. This is why, although their criticism was extremely harsh, they never used any anti-Semitic articles or even statements that could be considered anti-Semitic. Actually they tried to avoid the Jewish question altogether and did not even mention the usual anti-Israeli statements in describing Middle Eastern policy. Still, in the newspaper "*Pravda*" we found several places where Abramovich[12] was mentioned rather negatively in the electoral campaign as a symbol of oligarchs. Actually, articles were aimed against the government for enabling the oligarchs to steal money from the common citizens, but in both cases only Abramovich was mentioned by name as a symbol.[13] In other material dealing with the round table discussion devoted to the 90th anniversary of the October Socialist Revolution, one of the participants used the term "prepaid historical lies of Svanidze".[14] Practically the same situation can be seen when reading "*Sovetskaia Rossiia*". The only anti-Semitic hints are presented in very emotional letters by readers. For example, I.V. Lepnin, in his letter entitled "*Vernite bogatstva domoi*" (Bring the fortune back home) again mentions "*abramovichi, berezovskie, smolenskie*", i.e. those Jewish oligarchs who robbed Russian working people.[15] An editorial statement appears before the letters, apologizing for having to edit letters in accordance with "Putin's electoral law". Though the hints are quite clear, nobody can accuse the Communists of anti-Semitism because of their form. Besides, we are to assume that such minor cases would not even be noteworthy, and in a way this is proof that anti-Semitism was barely used in their campaign.

The same harsh criticism of Putin's administration, of the way Dmitrii Medvedev was "nominated" as his successor, of his refusal to participate in public debates, etc., was presented on the pages of major Communist newspapers during the President's electoral campaign. And again, neither during the period before the elections nor after them were there any attempts to use any anti-Semitic rhetoric.

In order to discover if there were differences in this trend on the regional level, we checked the newspapers of one of the regions with former serious Communist representation – *Ul'ianovskaia guberniia*. Out of more than twenty existing regional newspapers we looked through four socio-political ones – "*Narodnaia gazeta*" (The People's Newspaper, published by the Administration of the region 3 times a week, 37,000 copies), "*Simbirskii kur'er*" (The Simbirsk[16] courier, an independent newspaper, published 3 times a week, 19000 copies), "*Levyi marsh*" (Left March, published by the Union of the Left forces twice a month, 10000 copies) and "*Ul'ianovskaia Pravda*" (Ul'ianovsk truth, published by the Administration and *Zakonodatel'noe sobranie* of the region, 4 times a week, 4600 copies). In none of them during the electoral campaigns did we find any examples of anti-Semitism.

Even the so-called "patriotic" newspapers did not directly use anti-Semitism during the elections. For example, *Spetsnaz Rossii* published only 3 items devoted to the elections to the State Duma. Two of them dealt with the newly elected former members of the Special Forces – Igor' Barinov (*Edinaia Rossiia*) and Oled Denisenko (Communists)[17] while the third, in an article entitled "*Kto kogo i kak vybiral*" (Who elected whom and how) by Andrei Bortsov,[18] presented a detailed critical analysis of the elections. In this article he accused the liberals of defending the interests of the minorities – businessmen and national clans - against the freedoms and possibilities of the Russian majority, and called them perverts, but did not specify whom exactly he meant. *Spetsnaz* did not react at all to the presidential elections, and abstained from any anti-Semitic statements or hints. Likewise the newspaper "*Patrioty Rossii*".

As for the state circulated newspapers,[19] none of them used any anti-Semitic statements during both electoral campaigns. This does not contradict their customary position, since they did not print such material during the entire monitored period (2006-2007).

We also especially checked newspapers published by Jewish communities, ("*Evreiskaia gazeta*", "*Evreiskie novosti*" and "*Mezhdunarodnaia evreiskaia gazeta*") and discovered that in general they barely mentioned the problems of the elections. Not only did these newspapers never show any electoral activity of Jewish voters or present any programs of Jewish candidates, it seemed as if the issue of the elections itself was not of primary importance in their agenda. For example, "*Evreiskaia gazeta*" published only one article by Il'ia Mil'shtein, "*Marsh prokhodimtsev*" (March of rogues), in which he wrote that practically nothing depends on the voting of common citizens and came to the conclusion of the impossibility of fair elections.[20] "*Evreiskie novosti*" published only one editorial piece, "*REK privetstvuet vydvizhenie Dmitriia Medvedeva*" (Russian Jewish Congress greets the advancement of Dmitrii Medvedev) in which it stated that the representatives of religious and national minorities of Russia highly value Dmitrii Medvedev's position regarding the struggle with Nazi propaganda, chauvinism and any forms of intolerance.[21] The most active in this respect, "*Mezhdunarodnaia evreiskaia gazeta*" published several items devoted to the elections, including interviews with candidates Zhores Alferov and Nikolai Gonchar[22] [both are non-Jewish, but I am not sure we are to stress it – BM][23] in which they expressed their political views on the eve of the elections; two articles devoted to the problems of sociological services producing the surveys of the population before the elections and their political importance.[24] The results of the Duma elections were presented in the article "*Vybory v Dumu vyigrali Putin i kommunisty*" (3 December 2007). Two items were published regarding the presidential elections: an interview with Evgeniia Albats[25] given on the radio station "*Ekho Moskvy*" (Echo of Moscow) (27 December 2007), and an editorial piece entitled "*Ziuganov ne otkazhetsia ot uchastiia v prezidentskikh vyborakh*" (Ziuganov will not reject participation in presidential elections) (1 February 2008). But in none of the above mentioned items were any specifically Jewish questions discussed.

We paid close attention to the newspaper "*Zavtra*". As we discovered during the monitoring period, "*Zavtra*" is probably the only newspaper with a comparatively large circulation (over 100000 copies), officially distributed via kiosks, that adopts statements and even articles containing strong anti-Semitic undertones.[26] "*Zavtra*" usually makes keen comments on all important political events in the country, and in a way the comparatively small number of articles devoted to the elections to the State Duma might indicate the lack of political struggle. In fact, only three articles really addressed the Duma elections. One of them was an editorial, "*Pravda i krivda sistemy 'Vybory'*" (The truth and lie of the system "Elections"), explaining the policy of the new Head of the Electoral committee Vladimir Churov, who declared that there would be no "black PR", no false bulletins or any other forgeries. The largest part of this material is an interview with Iurii Lomov, the author of the electronic system for counting votes, who explained the impossibility of falsifying the results.[27] Another article by Artur Beliaev, "*Angel'skie vybory*" (Angel's elections) showed the absolute absence of any political struggle during the election, when everything is decided in advance.[28] The results of the elections were discussed by Aleksandr Nagornyi and Nikolai Kon'kov in the article "Prodolzhenie proidennogo" (The continuation of what has passed).[29] Aleksandr Prokhanov reacted to the elections to the Duma with two articles, "*Imperskii marsh izbiratelei*" (Imperial march of voters) and "Mnogo *zvanykh, no malo izbrannykh*" (Many called but few were elected),[30] but the main idea of both of them is the necessity of a strong, one party political system in the country and the end of the "epidemic of liberalism". It is very interesting to note that during the entire electoral campaign "*Zavtra*" published only two anti-Jewish hints. One has already been mentioned in the article by Artur Beliaev, where he wrote that a possible scandal before the Duma elections could only occur in the event Chubais would be caught by paparazzi in church with a candle. Another was contained in a poem by Evgenii Nefedov, in which he wrote that "*chubaises, shvydkois*[31] and others who are protected by the authorities have now become the builders of new Russia".[32]

The presidential elections, however, attracted many articles and items in "*Zavtra*".[33] As usual, practically all of them propagated "*Zavtra's*" traditional stance regarding the necessity of a third term for Vladimir Putin. This concept was openly presented by Aleksandr Prokhanov in his editorial article, "*Ot politicheskogo lidera k dukhovnomu vozhdiu*" (From political head to spiritual leader), in which he declared that people considered Putin the man who had saved Russia from disintegration, the "*gosudarstvennik*" (Man of State) for whom power is just the means to realize his Messianic mission to return Russia its eternal values and to reconcile people with God, from whom the liberals have tried to cut them off. Putin was named in a row of such Russian state symbols as Aleksandr Nevskii, Dmitrii Donskoi, Peter the Great and Stalin.[34] The results of the presidential elections were summed up by Prokhanov in his editorial, "*Kogda rasseialsia predvybornyi tuman*" (When the pre-electoral mist evaporated) in which the author declared that the transition of power was successful and consequently the idea of Russian development as a strong, unique state was continued.[35] It is quite significant that "*Zavtra*" tried to do its best to be tolerant and politically correct during the entire election period, and avoid any direct anti-Semitic (or other xenophobic) statements.

In general, we concede that during the monitoring of state-circulated Russian newspapers, as well as of newspapers of political parties that participated in the elections or their pre-electoral material published in the regular newspapers,[36] there were no anti-Semitic statements, slogans, articles or any other expressions.

This fully corresponds with the statement of the Chief Rabbi of Russia, Berl Lazar, who in his interview said that xenophobic slogans were not used during the electoral campaign and that none of the participants of the political process played the so-called "national card". "Today it has just stopped being profitable. Politicians understand that it brings them more problems than benefits. I consider this the main achievement of the recent campaign." – said Berl Lazar.[37]

Anti-Semitism during the electoral campaigns

Berl Lazar was only partly right. Unlike the newspapers of the parties participating in the political struggle for power, extremist and radical newspapers of different kinds still employed their customary anti-Semitic and xenophobic propaganda.

The largest of the radical newspapers that still exist, "*Duel'*" (Duel) paid scarce attention to the elections to the State Duma. It seemed as if they foresaw no political changes as a result of the forthcoming elections. The only pertinent piece, "*Rasstreliat' ili izvinit'sia*?" (To shoot or apologize?)

by V.Bushin levels harsh criticism at the Communists, who on the eve of the elections expelled several members of the CPRF, including Albert Makashev, well known for his rampant anti-Semitism, from the party. The author wrote that showing such ruthlessness towards their own friends, the Plenum of CPRF, seriously helped Svanidze, Mlechin and Radzinskii[38] to create the slanderous image of the bloodthirsty Bolshevik.[39]

The presidential elections, however, did attract their warped attention. On February 26, 2008, *"Duel"* published an unsigned piece reprinted from *pravda.ru*, "*Zato u nas evrei stanet prezidentom*!" (In return a Jew will become President here!). With no connection to the title whatsoever, the published item contains data on the loss of power of Russia in the modern world, decrease in the population of the country, etc.[40] In the same issue, another article appeared by Ruslan Bogatyrev, "*Den' pozora*" (The day of disgrace.[41]) The article opens with a caricature of a clearly anti-Semitic character, showing the change of head (Putin's for Medvedev's) of a monstrous figure formed by Jewish hands. In this article the author states that formerly mighty Russia now humbly lies down under the heel of Zionism. Very soon Russia will become a protectorate of Israel. "Why does Russian Putin, - asks Bogatyrev - hand over the fate of Russia to the Jew Medvedev, who will be a puppet of the oligarchs, Berl Lazar, Israel and the USA?" The author also insists that the appointment of Medvedev as successor was carried out after the meeting between Putin and Abramovich, in which Abramovich presented an ultimatum. A brief biography of Medvedev was also mentioned, especially stressing the fact that his mother, Iuliia Veniaminovna, was Jewish. The article closes by declaring that the new Head of Russia is a henchman of criminal oligarchs, the Federation of Jewish Communities of Russia and a potential citizen of Israel.

A similar article, "*Vazhen ne sam vybor, a ego posledstviia*" (What is important is not just the choice but its outcome), signed by one Victor

Innokent'evich (patronymic), presents modern Russia as a country undermined by liberals seeking to appropriate all its riches. The group comprising Chubais, Medvedev, Kudrin, Gaidar, Voloshin, Abramovich, Miller and Gref[42] had a plan of seizing presidential power, which was successfully realized. They are compared with Khazars – "the first serious enemies of the Russian state". But the author does not want to leave his reader with any doubts, and thus also cites a quotation from an article by Moscow Rabbi Pinkhas Goldshmidt:[43] "Today Jews have power, money and influence in Russia like never before". Finally, he concludes that the election of Medvedev will close the circle and he will be under the control of Zion. Naturally, this new idea is presented together with familiar ones of the domination of Jews in the State Duma, TV, etc.[44]

These articles demonstrate a somewhat new tendency of presenting Dmitrii Medvedev as the Jew, henchman of Zionists and Israel, which was widely used, albeit with minor changes, by many other radical and extremist newspapers.

The newspaper "*Potaennoe*" (Hidden)[45], for example, states that the appointment of Medvedev was a direct reaction from the Jewish world to the open islamization of Europe and the USA. As a result, Russia will become a country with a prepared Jewish infrastructure and many Jews will return to Russia from Israel and other countries. Medvedev became a very favorable candidate for the Jewish media all over the world not only because of his partly Jewish origin but also because of his active position in struggling with nationalism and anti-Semitism.[46] The same issue published the so-called apocryphal biography of Medvedev, which stated that although he is Russian according to his passport, his mother is Jewish. It also stressed that his wife's maiden name was Jewish – Linnik. Another unsigned article "*Antesimity Rossii razocharovany vyborom Putina*" (Anti-Semites of Russia are disappointed by Putin's choice) was also published in the same issue. It just states that though nothing bad could be said about Medvedev's mother, her Jewish nationality is known and that this will make the half-Jewish candidate unpopular with Russian voters, who in general don't like Jews and other "strangers".[47]

Immediately after the elections "*Potaennoe*" published another unsigned article, "*Glavnyi ravvin Rossii vystupil v Oksforde v obshchestve Shabad s lektsiei o evreiakh Rossii i ikh budushchem*" (Chief Rabbi of Russia delivers a lecture in Oxford regarding the Jews of Russia and their future).[48] The article quotes the speech of Berl Lazar delivered on 27 February 2007, in which he spoke about Putin's good attitude towards the Jews and said that no one ruler of Russia or USSR has done more for the Jews than Putin. He said that though there were many rumors in Russia that Medvedev was Jewish, Jews actually did not consider him to be Jewish, but three days before his nomination as successor Medvedev came to the Jewish center and promised that everything would go well for the Jews.

"*Russkii vestnik*" (Russian Herald) newspaper, published with the assistance and protection of the Russian Orthodox church, unlike many others was rather active on the eve of the elections to the State Duma, and among other items published several articles containing anti-Semitic statements. One of them was titled "*Nezadolgo do vyborov*" (Not long before the elections) and was written by A.V.Shcherbakov. The article actually stresses the problem of the Jewish role in the October Revolution of 1917. The author repeats well known and often used anti-Semitic myths stating that the Jews are to be blamed for the revolution, that practically all the Jews supported it and that the so-called "red terror" was declared by the Jew Sverdlov as a response to the mythical "white terror". He again accuses the Jews of serving in ChK and killing Russians and denies the existence of state anti-Semitism in the Soviet Union, claiming that the highest percent of Communists were Jews, etc. After such an impressive introduction, without any recognizable, logical connection, the author concludes that during the Duma elections of 2007 only the political party oriented to the national interests of Russians could become a real alternative to *Edinaia Rossiia*. But there is no such party, because the authorities had done their utmost to prevent its emergence. Finally, the author foresees a forthcoming crisis during the summer of 2008 and warns that it could end in a new popular uprising.[49] In the same issue, another article was published by Sergei Petrov, which was devoted to the decision of the Supreme Court regarding the disbarment of the "*Narodnyi Soiuz*" party (People's Union, led by S.N.Baburin) from participation in the Duma elections. The reason is obvious – this party defends the fundamental values of Russian society and hence presents a threat to the existing anti-Russian regime.[50] The position of the newspaper supporting *Narodnyi Soiuz* became even clearer when it published several articles accusing the Communist party and Ziuganov of distorting the real history of Russia.[51] Incidentally, one of such articles – "*Pochemu ia golosoval protiv demokratov i kommunistov*" (Why I voted against the democrats and communists) by A.N.Ignat'ev, contains such a statement: "About democrats people openly say that it is an assemblage of persons working against Russia in the interests of the USA and Israel. The ideology that unites them – is Zionism." Criticizing the Communists, the author states that they repeat Jewish lies, popular after the revolution, about the illiteracy of Russians before 1917.[52]

As for the presidential elections, "*Russkii vestnik*" used already mentioned items regarding the apocryphal biography of Medvedev,[53] mentioning his Jewish mother and Berl Lazar's lecture in Oxford,[54] as well as some new pieces. Among them was an article by Filipp Lebed, "Haim Ben-Iaakov: "*U nas s Rossiei nastol'ko raduzhnye otnosheniia, chto dazhe strashno*" (Haim Ben-Iaakov: We have such great relations with Russia that it is even frightening) which presents an interview of the Representative of the Jewish Agency in Russia. Quotations of Ben-Iaakov are used to prove that there are no manifestations of anti-Semitism in today's Russia, and the author also emphasizes the already mentioned fact of Medvedev's visit to the Jewish center.[55] Another article, "*Evreiskie mal'chiki*" (Jewish boys) by Arkadii Krasil'shchikov, is an abridged version of an article reprinted from "*Evreiskoe slovo*" (The Jewish word).[56] In this article its author, an Israeli journalist well known in the struggle with anti-Semitism, expresses his fear that active participation of Jewish youngsters involved in Russian politics could result in new anti-Semitic pogroms. This is why he even votes for the establishment of a so-called "percentage norm" for Jews in the Russian political sphere, meaning that if for example Jews number 1% of the total population of Russia, their representation in the State Duma and political structures should also not exceed 1%. The author addresses young "revolutionaries of Semitic appearance": "You are talented, but you immediately become strangers when you perch on the dirty minefield of political discord." But this article, published in "*Russkii vestnik*" in a certain context, produces the impression of an Israeli accusing Jews of attempts to rule in Russia, and has the opposite effect to what was intended. The results of the presidential elections were commented upon by the same A.V.Shcherbakov in his article, "*Nachalo puti k 2020*" (The beginning of a way to 2020), in which he writes: "I hardly think that during Medvedev's period in Russia something will change in favor of the Russian people".[57]

Another anti-Semitic newspaper, "*Pravyi vzgliad*" (Right view),[58] also reacted to the presidential elections. In the article by A.V.Krainev, "*Posadili na* 'tsarstvo'"[59] (Seated for the 'reign'), the author states that the representative of one of the Jewish tribes will now rule Russia for the next 4 years. He writes that what Andropov[60] didn't manage to do would now be done by Medvedev. He also expresses serious concern that the rather active wife of the newly elected president, Svetlana Medvedeva, will become a second Raisa Gorbacheva and will actively participate in the process of ruling the country under the influence of certain Zionists like Pozner. An interesting detail: the article is illustrated by a drawing by Dmitrii Medvedev bearing the slogan, "*Lzhedmitrii – von iz Kremlia*!" (False Dmitrii – out of the Kremlin!).[61] In the same issue a photo of Medvedev together with Berl Lazar appears with a corresponding article, "*Khasidy ochen' rady*" (Hasids are very glad). In this article the Berl Lazar interview is quoted, where he said that Russian Jews wanted to continue the very good status quo because they wanted to live peacefully during the coming years. An editorial comment to this item states, "Israel was the first to greet Menakhem Natanovich Mishkin and express its deep satisfaction with the results of the elections".[62]

In another issue of *"Pravyi vzgliad*" another article appeared preserving parallels between the false Prince Dmitrii and Dmitrii Medvedev (who is actually called Menakhem Anatol'evich).[63] This article states that both the false Dmitrii and Medvedev are impostors, that the false Dmitrii II was also Jewish, that they were very small (like Medvedev) and concludes that Russians don't like impostors.

Another newspaper, (Neva's Boundary)[64] published an article entitled "*Prokol*" (Puncture) by Valentin Kashinov, which states that although in many countries there are no laws that potential citizens of other states cannot be appointed to positions dealing with secrecy, no Jew in Russia can be appointed to a high state position (not to mention the presidency). But the Russian constitution gives such rights, and because of this the Jewish global government has managed to seize power in Russia in order to rob its native population and carry out "poli-ethnic genocide".[65] In the same issue, an article by Victor Uverov, "*On takoi zhe russkii natsionalist, kak Vladimir Vladimirovich. Kto by somnevalsia*" (He is the same Russian nationalist as Vladimir Vladimirovich. Who would doubt it) speculates on the quotation of Vladimir Putin from his speech during his meeting with Angela Merkel, when he said that Medvedev is in a good way the same kind of Russian nationalist as he himself. The author surmises that this was Putin's answer to the question that was probably asked, whether the ethnic Jew Menakhem as Russian president would adequately reflect the interests of Russians.[66] The same issue also presents the above mentioned lecture delivered by Berl Lazar in Oxford under the title, "*Berl Lazar: golovokruzhenie ot uspekhov*" (Berl Lazar: dizzy with success).

Of course we did not manage to find and examine all the existing extremist newspapers, but the above material provides sufficient grounds to draw some conclusions.

Conclusions

1. In general it is possible to say that the 2007-2008 campaigns were for the most part "clean" of anti-Semitism. No political party that participated in the Duma elections openly used any anti-Semitic statements in their propaganda. During the presidential elections none of the candidates (as well as parties supporting them) used any statements that could be considered anti-Semitic.

2. The so called "mainstream" newspapers or newspapers with national circulation were also politically correct, and never used xenophobic or anti-Semitic remarks or statements in presenting the political platforms of the different political parties during the election period.

3. Nevertheless, open accusations against Jews, anti-Semitic articles

and statements were found in the marginal extremist press. These were especially prolific during the presidential elections, and the majority of them were devoted to the supposedly Jewish roots of the newly elected president, Dmitrii Medvedev.

Notes

1 There are several *gubernias* of the so called Communist or "red" zone, where the majority in the local Legislative Councils are kept by Communists and there are still Communist Governors.

2 *"Edinaia Rossiia"* (United Russia) – leading party keeping majority in the State Duma of Russian Federation. Was created in 2001 by uniting of 2 organizations *"Edinstvo"* (Unity) and *"Otechestvo"*(Motherland) into one joint political party. During presidential elections of 2008 positioned itself as a pro-presidential party and proposed the candidature of Medvedev.

3 SOVA – analytical center in Russia monitoring manifestations of racism and xenophobia.

4 This figure looks more reliable and was independently proved by the research of well-known Israeli demographer Mark Tolts.

5 The vast majority of Jewish population of Russian Federation now lives in Moscow, S-Petersburg and large cities.

6 Ziuganov, Gennadii Andreevich (born in 1944) – the Chairman of the Central Committee of the Communist Party of Russian Federation (CPRF) since 1995. During the presidential elections of 1996, 2000 and 2008 was nominated as a candidate of the Communist Party.

7 Document: http://www.regnum.ru/news/925633.html - Главный раввин Берл Лазар: "Результат выборов очень важен для евреев России" (Chief Rabbi Berl Lazar: "The result of the elections is very important for Russian Jews"), published 20:13 02.12.2007.

8 During 2006-2007 several newspapers expressing anti-Semitic and xenophobic manifestations were banned according the court's decisions: *"Zyrianskaia zhizn'"* (The life of Zyrian') was closed in 2006, *"Rus' pravoslavnaia"* (Orthodox Russia), *"Russkoe veche"* (Russian veche), *"Russkii front"* (Russian front), *"Istina"* (Truth), *"Krasnoiarskii komsomolets"* (Member of Young Union League from Krasnoiarsk) and some others - in 2007. During the same period several chief editors of these newspapers were condemned to 2-3 years (conditionally).

9 Examples of usage of the nationalistic slogans and anti-Semitic rhetoric could be seen during the elections to State Duma in 1999 in the propagandistic materials of LDPR; in xenophobic propaganda of the party "Rodina" (Motherland), anti-Semitic statements of one of the Heads of the Communist party General Albert Makashov and nationalistic slogans of one of the leaders of "Rodina" Sergei Glaz'ev during the elections of 2003.

10 See his newspaper "LDPR", 29 October 2007, in which he states that there are so many of them in Moscow that one can hardly hear Russian language, accuses them of carrying out the vast majority of all crimes in Russia, etc.

11 Using this word always meant Jews – people selected by God. See Newspaper "LDPR", 1 October 2007.

12 Abramovich, Roman Arkad'evich (born in 1966) - one of the richest Russian businessmen, lives in London.

13 See, an article by Igor' Mel'nikov "Kvartira – v Londone, den'gi – v Shveitsarii" (The flat - in London, money – in Swiss) – Pravda N 126, 15 November 2007; editorial "Za chto nam blagodarit' Putina" (What for we are to thank Putin), in which it is stated that Putin established the same tax (13%) for the fat cats and for the poorest people – "It is the same for Abramovich and the cleaning lady" – Pravda N 130, 23-26 November 2007.

14 Svanidze Nikolai Karlovich (born in 1955) – well known Russian journalist and TV producer. Conducts several popular programs and among them "Historical chronicles"- Pravda N 122, 2-7 November 2007.

15 Sovetskaia Rossiia N 21, 28 February 2008.

16 Simbirsk – pre-revolutionary name of the city Ul'ianovsk.

17 Spetsnaz Rossii N 12 (135), December 2007, articles by Pavel Evdokimov "Nashego polku pribylo" (There are more of ours) and "Polpred spetsnaza -2" (Charge d'affairs of special forces -2).

18 Ibid.

19 We checked the newspapers of liberal orientation ("Novaia gazeta", "Gazeta"), newspapers representing official point of view ("Vedomosti", "Rossiiskaia gazeta", "Nezavisimaia gazeta", "Izvestiia", etc.), tabloids ("Zhizn'", "Komsomol'skaia pravda", "Moskovskii komsomolets", "Argumenty i fakty").

20 Evreiskaia gazeta N 12 (64), December 2007.

21 Evreiskie novosti N 41 (227), December 2007.

22 Zhores Alferov (born in 1930) – famous Russian physicist, Noble prize winner (2000), member of State Duma since 1995, since 1999 – represents there Communist Party; Nikolai Gonchar (born in 1946) – Russian economist and politician, member of State Duma since 1993, now represents "Edinaia Rossiia".

23 Mezhdunarodnaia evreiskaia gazeta, 24-25 October 2007.

24 Article "Sotsiologi vrut?" (Do the sociologists lie?) republished from "Izvestiia" 26 November 2007, and "Liubliu, no ne zhenius'"(I love but will not marry) – interview with B.Dubin (29 November 2007).

25 Evgeniia Al'bats (born in 1958) – well known Russian journalist.

26 See detailed review in our report.

27 Zavtra N 42 (726) 17 October 2007.

28 Zavtra N48 (732) 28 November 2007.

29 Zavtra N 49 (733) 5 Decamber 2007.

30 Zavtra NN 41 (725) 10 October 2007 and 49 (733) 5 December 2007.

31 Using of these surnames of people considered to be symbols of Jewish evil influence in politics, culture and TV without capitalizing in plural is very typical for anti-Semites.

32 Zavtra N 48 (732) 28 November 2007, "Evgenii o nekikh" (Evgenii regarding certain people).

33 See for example, Aleksei Podberezkin "Uiti chtoby ostat'sia" (To leave in order to stay) – Zavtra N 40 (724) 3 October 2007; Vladimir Arkhangel'skii "Putin– ne aiatola" (Putin is not an aiatola); Vladimir Bondarenko "Sud'ba 'zlodeia'" (The fate of a "villain"); Vladislav Smolentsev "Kremlevskii defolt" (Kremlin default) – Zavtra N 46 (730) 14 November 2007; Vladimir Arkhangel'skii "Upravliaemyi absurd" (Directed absurd) – Zavtra N 7 (743) 12 February 2008; Aleksandr Prokhanov "Putin - Medvedev: peresadka organov" (Putin – Medvedev" transplantation of body organs) – Zavtra N 8 (744) 20 February 2008; etc.

34 Zavtra N 43 (727) 24 October 2007.

35 Zavtra N 10 (746) 5 March 2008.

36 The vast majority of the parties didn't have newspapers of their own.

37 See, http://www.regnum.ru/news/925633.html - published 20:13 02.12.2007.

38 TV stars of Jewish origin well known for their popular historical programs.

39 Duel' N 49, 4 December 2007.

40 Duel' N 9, 26 February 2008.

41 Ibid.

42 This row is a very strange combination of Russian economists who headed the transition of Russian economy from the socialism to capitalism (Chubais, Gaidar, Voloshin) with some people who appeared in the government as ministers of finance and economics (Kudrin, Gref) during the Putin's times and oligarchs (Abramovich, Miller). The only factor characterizing all of them is their non-Russian (or supposedly non-Russian) origin.

43 Refering it to Ha-arets, 15 December 2005.

44 Duel' N 11, 11 March 2008.

45 Published in S-Petersburg by R.L.Perin with announced circulation of 5000 copies.

46 See an article by R.Khorsov "Islamskii factor pri vybore preemnika" (Islamic factor in the process of choosing the successor) – Potaennoe N 1(38) 2008.

47 Potaennoe N 1 (38) 2008.

48 Potaennoe N 3 (40) 2008.

49 Russkii vestnik, 23 November 2007.

50 "Ob otstranenii Narodnogo Soiuza ot vyborov v Gosudarstvennuiu Dumu" (Re debar of Narodnyi Soiuz from the elections to the State Duma) – Russkii vestnik, 23 November 2007.

51 See for example an article by V.Shcherbakov "Novyi etap raspada Rossii nachinaetsia?" (Does the new stage of disintegration of Russia start?) – Russkii vestnik 8 November 2007.

52 Russkii vestnik, 6 December 2007.

53 "Iz apokrificheskoi biografii D.A.Medvedeva' (From the apocrypha biography of D.A.Medvedev) - Russkii vestnik, 21 December 2007.

54 "Glavnyi ravvin Rossii Berl Lazar o V.V.Putine I perspektivakh evreev v Rossii" (Chief Rabbi Berl Lazar about V.V.Putin and perspectives of Lews in Russia) – Russkii vestnik, 13 March 2008.

55 Russkii vestnik, 21 December 2007.

56 The given reference is for "Evreiskoe slovo" N 48 (370) 2007 – Russkii vestnik, 15 February 2008.

57 Russkii vestnik, 11 April 2008.

58 The newspaper of Russian Imperial Movement, published in S- Petersburg (998 copies).

59 Pravyi vzgliad N 5 (57), March 2008.

60 There were rumors and speculations that Iurii Andropov (Secretary General of the CPSU) was also of Jewish origin.

61 Direct hint for the henchman of Polish kings – false Prince Dmitrii (actually one of many), who managed to conquer Moscow and later was killed by Russian people in Kremlin.

62 Here the name, patronymic and surname of Medvedev are twisted (or translated) to stress his Jewishness.

63 See an unsigned article "Pro samozvantsa" (About Impostor) – Pravyi vzgliad N 6 (58), March 2008.

64 The newspaper of Nevskii district of S-Petersburg, (999 copies).

65 Nevskii rubezh N 15, April 2008.

66 Ibid.

Yehuda Bauer

Holocaust Research – A Personal Declaration

All studies of the Holocaust face one basic question that people only rarely consider on a conscious level: Was the Holocaust an historical event that can be analyzed and understood no less, but also perhaps no more, than any other historical event, or was it an event that defies explanation and one that humans are simply incapable of understanding and internalizing? That is to say, might the Holocaust as a specific series of historical events possess a fundamental essence that transcends all human comprehension?

As one of many scholars of the genocide that National Socialist Germany and its collaborators throughout Europe perpetrated against the Jewish people, which we refer to as the Holocaust, I have a responsibility toward my readers and toward myself to answer this basic question. My answer is that the Holocaust was an event, or a series of events, that was carried out by human beings based on human reasons and human motivations, and that neither God nor Satan (in the event that they exist, and I happen to believe that they do not) had anything to do with it. As such, in principle although not in practice, the Holocaust can be understood to the same extent that we can understand the history of the Pilsudski regime, British policy in India, or the French Revolution. Anything done by people can be understood by other people due to the similar (though by no means identical) nature of the motivations, contexts, and characteristics of the figures involved. The problem, then, is not that the Nazis were *inhuman,* but rather that they were *human.* Within each one of us there is a kernel of Himmler and Eichmann which, given the appropriate circumstances–in a different social, political, or cultural context, or perhaps if we had a different family background–could potentially develop into mass criminality. There is nothing mystical about the Holocaust, which was a human act of atrocity of a most extreme nature. In this way, it must be understood as constituting the extreme edge of something that does indeed exist in human history and that can therefore be explained and understood. This does not mean that we actually understand it in practice or that there is any possibility of our understanding it fully in the near future. However, contrary to post-modern assertions, I posit that the Holocaust was the outcome of an endless number of causal chains whose centrality we ultimately have a good chance of understanding, and many others about which we can make educated guesses.

But the $64,000 question, as some Americans like to put it, is this: Whence came this compulsion to commit mass murder in general and the mass extermination of certain human groups in particular? The real answer, of course, is that I do not know. I doubt that anyone can provide an answer to this question that is scientifically provable. My conjecture, however, is that we all possess the instinct for mass murder. After all, all humans on earth–Hitler, Stalin, Einstein, the Australian Aborigines, and you and I–are descendants of a group of Homo sapiens (the ostensibly intelligent modern man) that wandered the savannas of East Africa some 120,000-200,000 years ago, give or take a few millennia. This much has been proven by DNA testing, confirming that separate races do not exist and that there is a grain of truth to the biblical myth of Adam and Eve, which intimates that we all belong to one human race of the same origin (and that also intimates–undoubtedly unintentionally, as all monotheistic religions are anti-feminist–that woman is the source of all knowledge, as she was the one who plucked the fruit from the tree of knowledge). We are territorial mammalian predators, like leopards, bears, and the like. However, we lack the physical constitution of big cats and bears, making our ability to hunt dependent on cooperation. Humans killed animals: they had to in order to survive. Today, they no longer hunt to kill and eat mammoths - rather, they shop in supermarkets and buy meat and fish off the shelves. Fundamentally, however, it is the same thing: except for the few vegetarians among us, we continue to exist by killing animals and fish. Every group that has invaded or that invades the real or virtual realm of another group, and threatens to damage the home group's real or virtual ability to conduct life as usual, is absorbed, assimilated, enslaved, expelled, or killed. These are the behavioral responses that can be observed from the onset of human history. In this way, not only killing, but killing in large numbers in the struggle over territory and hunting and gathering areas, was once an essential human need. Over the millennia, it developed into something akin to instinct that still exists within us today.

If there is any truth to this anthropological conjecture and if the instinct for genocide is truly inherent within us, this raises the question of whether it is at all possible to prevent acts of mass atrocity or mass murder. If not, then all our political and educational efforts to do so have been in vain. However, we can find comfort in a different element of human instinctive behavior – the fact that humans also needed to cooperate with one another to hunt, gather fruit and so on, and that in order to fulfill this need, they developed emotional sentiments such as caring, love, sympathy, and the instinct to save human life – even that of someone we have never met. This behavior may be based on the subconscious assumption that a person whose life we save becomes a potential ally and a partner in the struggle for existence. Evidence of this instinct has emerged in many cases of genocide: indeed, according to extremely strict criteria, Yad Vashem has recognized more than 21,000 Righteous Among the Nations. My assumption is that this number represents only a percentage, and perhaps only a small percentage, of those who actually saved lives. Countless others were undoubtedly killed along with those they were trying to save, and in some cases the people they saved may have never even known their names or did not live to tell their stories. In other cases, the names of those who saved lives were simply forgotten. There were also some populations of which a majority either saved or supported efforts to save Jews, such as the Danes, the Bulgarians, who saved Jews in Bulgaria itself (although their government handed over Jews found in areas under Bulgarian control, in Thrace and Macedonia, to the Germans), many Italians (although there were many anti-Semitic fascists and traitors among them), and Serbs. When considered together, these individuals still represent only a small minority. The overwhelming majority of Europeans who lived under the rule of the Nazis or their allies were either hostile to the Jews or viewed the fate of their Jewish neighbors with cold indifference, and this minority reflects the other side of human behavior, if it can be referred to in this manner, under such difficult conditions. Their very existence suggests that the struggle against recurring acts of genocide, though certainly an extremely difficult battle, is not a hopeless one.

The Holocaust was undoubtedly a form of genocide. The 1948 UN Convention on the Prevention and Punishment of the Crime of Genocide was instigated in the United States by attorney Raphael Lemkin, a Jewish-Polish refugee, based on what he knew in 1943-44 about the mass murder of the Jews and the Poles (although he proposed something similar one decade earlier, in 1933, in a lecture he sent to a convention of attorneys engaged in problems relating to international law). But do we accept the definition of genocide advanced in the Convention? As it is truly inadequate, I know of no serious academics who do. All of our historical and social scientific definitions are invented social and historical abstractions, and such realities are always much more complex than the definitions we assign to them. Moreover, many of us faced with these complexities attempt to adapt reality to fit the definitions, instead of doing the opposite. As a result, I and many of my colleagues have done our utmost to abandon the useless debate over definitions and have instead started seeking alternative approaches. In my view, the best option is to rely on descriptions instead of definitions. When we do so, we can say that acts of genocide, massacres that are genocidal in nature, and acts of mass cruelty (the terms are interchangeable) are perpetrated against real or virtual groups. This includes "politicide," a term introduced by Barbara Harff to refer to the real or virtual mass murder of groups of a political, economic, or social nature. The Kulaks, for example, who were subject to "extermination as a class," in the words of Stalin at the onset of collectivization in the Soviet Union in 1929-30, constituted not a real group but an imaginary one. However, it was precisely classification in this manner that transformed them into a group that was very real, and people who were identified as Kulaks became the target of mass acts of atrocity. The same can be said of the Tutsi and Hutu in Rwanda, who were not truly two different ethnic groups, as they spoke the same language, were loyal to the same Christian communities, and shared a common culture. The distinction between these two groups developed on the basis of social class through a process intensified by the German and Belgian colonialists who had taken over Rwanda in the service of their rule. The moment they were defined as distinct groups, they became what can be referred to as "virtual ethnic groups." What these two tragic genocides have in common is the intention and the practical act of attempting to exterminate the groups in question, as described by the Covenant, by means that included mass murder (and other criminal acts). Other scholars have also proposed including "cultural genocide" in the definition of genocide. This results in a terminological contradiction, as the term "genocide" means "the murder of a people." A group deprived of their cultural heritage has a chance of rehabilitating itself on a cultural level. People that are murdered, however, cannot rehabilitate their culture.

In the meantime, historian Ben Kiernan published the first book on the history of genocide (*Blood and Soil: A World History of Genocide*, Yale University Press, 2007) which aroused significant debate. Research on this subject is essential, as acts of mass atrocity intended to exterminate groups as groups have been taking place since the dawn of humankind. They differ from one another in context, method, means, motive, and outcome. But even if my previous contentions are only roughly accurate, we nonetheless observe a common basis for all these human acts. Aside from this basis, what do all acts of genocide, including the Holocaust, have in common? First of all, we can point to the suffering of the victims, which in all these acts of mass atrocity must be regarded as equal. There is no scale for victimization, and no genocide is better or worse than any other. There is no torture, no murder of children, and no rape that is better or worse than other crimes of the same nature. Assertions along the lines of "we suffered more than you" are an absolute distortion of real individual and shared experiences, although they can certainly be understood in psychological terms. Another parallel between genocides is the fact that the aggressors always make use of the "best" means at their disposal to carry out their scheme. The Germans had gas, a first class bureaucracy, military might, and advanced technology, and they used them. The Hutu Power attackers in Rwanda had no gas, but they had a very good bureaucracy, developed by the German and Belgian colonialists on the foundation provided by the pre-colonial kingdom, which was astoundingly well organized. They used radio to incite and issue calls for murder, and they were able to identify the Tutsi everywhere using their identification cards. They were able to instruct the local Hutu in all cities and villages with regard to who to kill. They used machetes, knives, clubs, and weapons. They set up special death squads, most prominently the *Interahamwe*, which were comparable to similar units used by the Ottoman Turks to murder the Armenians, and to the special S.S. murder squads who served the Nazis.

Nonetheless, the means used by the Germans were unprecedented: for the first time in history, factories were built for the production of dead bodies. And although in Birkenau, Treblinka, and Chelmno thousands of Roma and a few hundred Polish and Russian prisoners of war were also murdered, ninety-nine percent of the victims were Jewish. Killing Jews was the ultimate aim of the establishment of these "death factories," a term coined by Erich Kulka and Ota Kraus in their 1957 book, *The Death Factory* (*Továrna Na Smrt*, Praha, NašeVojsko).

There are of course differences between genocides, particularly with regard to background, motivating factors, and explanations. However, I posit that no one case of genocide, except the Holocaust, possessed unique elements that were not present in other cases. For example, in the case of the genocide currently taking place in Darfur, we can speak of a number of causes, such as the African tribal groups' long history of dissatisfaction with the central government in Khartoum, against the background of the diminishing local natural resources on the one hand and the sharp increase in national income in Sudan stemming from the discovery of oil on the other hand. The dramatic rise in government income has been channeled into the development and enrichment of the Arab tribes of northern Sudan, which make up the majority of the government and social elite. The government itself is a radical Islamic government which has already committed genocide against the tribes of the black animist Christians in the south and therefore already has experience in using local militias to cover up acts of murder, even though they enjoyed the massive support of the official military. The ideology of the local Bedouin in Darfur, which constitute the backbone of the government-supported militias, is explicitly racist and calls for eradicating the black tribes cultivating land in Darfur. The same factors, including ideological justification, were also at play in other cases of genocide that have taken place over the past 150 years. In Darfur (as elsewhere), the motivations were financial, social, and, in part, anti-black ethno-racist. The perpetrators were primarily Bedouin Arab Aballa militias known as Janjaweed. Another example is the mixture of ethnic, military, and economic motivations that characterized the Hutu's genocide of the Tutsi in Rwanda, although their main motivation was political: that is, preservation of the exclusive political power of the Hutu elite, which originated primarily from northwestern Rwanda. These groups, who were had never been part of the pre-colonial Tutsi monarchy, continued their political struggle against the Tutsi by genocide. The Tutsi military threat was very real, as a Tutsi army had invaded Rwanda in 1990 to topple the Hutu dictatorship. A real threat also emerged in Darfur, when the black tribes rebelled against the government in Khartoum in 2003. In this case, the uprising was well organized, although by the time of writing the rebels had fragmented and lost much of their strength. All of these developments have parallels and precedents in a variety of genocides and other similar events.

German policy toward the Poles was also a policy of genocide: the intention of the Germans was to eliminate the Polish people as a national group (but not to murder the majority of the people who comprised it). To this end, they murdered as many members of the Polish elite as possible, eliminated or seized all Polish economic institutions, and totally destroyed all Polish institutions of middle and higher education. Their motivation was political and economic: they aspired to turn the Poles into slaves that would work for the benefit of National Socialist Germany as an element of Germany's rule over *Lebensraum*, its Eastern European "living space." In an effort to implement this policy, the Germans murdered about two million Poles. And again, we find parallels with other genocides – an amalgam of themes of political, economic, and strategic expansion that led to the perpetration of murderous atrocities. Unfortunately, this is nothing new.

Some of the same factors were involved in the Holocaust, including a powerful central force with a murderous ideology and military superiority (which was true also of the Ottoman's campaign against the Armenians), a state of war (as was the case in Rwanda, *mutatis mutandis*, of course), the robbing of the victims before and after their murder, and the economic use of Jews as slave labor prior to extermination. However, the Holocaust was also characterized by acts and motivations that were completely unprecedented. The chief one was the Nazis' goal of identifying and murdering all people they classified as Jews, without exception. In stages, they developed this aspiration into the also unprecedented aim of killing all Jews the world over, thereby committing genocide on a global scale, as Hitler explicitly explained to the Mufti of Jerusalem, Haj Amin al-Husseini, on November 28, 1941. Another unique aspect of the Holocaust with respect to the examples offered above and the other cases of genocide with which I am familiar was the fact that Nazi antisemitic ideology was entirely non-pragmatic. Although they robbed the Jews, the acquisition of Jewish property was not the motive for murder. Rather, the robbery was a "natural" byproduct of the passion first to expel all Jews and then to murder them. Nazi ideology was not based on economic calculations or considerations of efficiency. Had capitalist calculations of profit and loss been more important to them than ideology, they would not have murdered the Jewish slaves in their service. They would not, for example, have liquidated the ghettos of Lodz and Bialystok against the requests of the Wehrmacht (the Germany army) and the local Nazi bureaucracies, as these ghettos were providing them with essential products. To the best of my knowledge, this was the first time in history that genocide was carried out first and foremost for ideological reasons. To use Marxist terminology (I am not a Marxist, although I once was), this super-structure was not a product of the relations of production. Rather, the super-structure itself was the determining force that functioned contrary to economic and political interests. To use Marxist terminology once again, the super-structure imposed itself on the relations of production, contrary to Marxist doctrine. In most instances during the Holocaust, the Germans murdered Jews in against their own interests, allowing us to conclude that in this context, ideology trumped pragmatic interests. This also appears to have been unprecedented in human history.

The Holocaust, therefore, may be classified as the most extreme case of genocide known today. In other words, while it constitutes one of many examples of the general phenomenon of genocide, it also contains elements that have no precedent, a small number of which are enumerated above. My use of the term "unprecedented" here is intended to cast the Holocaust as a precedent with a potential for recurrence which in some ways has already repeated itself (for example, in the intention of the Hutu perpetrators of genocide in Rwanda to kill all those in the country they classified as Tutsi). On this basis, I argue that when we deal with genocide, we have no choice but to treat the Holocaust as an extreme paradigmatic event that may herald similarly extreme cases of genocide in the future. This is certainly true regarding efforts to prevent, or to at least reduce the chances of genocides in the future. The Holocaust is like a Sword of Damocles dangling above humanity, and all those seeking to engage the subject of genocide have no choice but to begin their exploration with this extreme, unprecedented example.

The Holocaust also has distinctly European aspects. It occurred in a society that was ostensibly democratic liberal, conservative, or social-democratic. It originated in a country that boasted some of the most advanced ideas and achievements in modern culture. A decisive majority of Europeans looked on from the sidelines, often rejoicing at the misfortune of the Jews or demonstrating hostile neutrality while their Jewish neighbors were taken away and murdered. In this way, they rejected their own ostensible religious and political ideologies. A large minority of the non-Jewish European populations took part in the killing and the looting that accompanied it. From a Christian perspective–and most Europeans were Christians, at least formally (that is to say, they had all been baptized)–the problem was even more serious. Some 1,900 years after the birth of the Christian messiah, members of his people were murdered by pagans who had been baptized into Christianity but who, through their actions or inaction, rejected the precepts of their religion, of which baptism was the major symbol. Most of the elite of the Catholic and Protestant churches chose to turn a blind eye. In some ways, it was a reenactment of Peter's denial of the messiah in the New Testament. Millions of roosters crowed, but very few listened.

Religious Jews were faced with a different problem. Where was God when his own People were being exterminated? How was it possible to explain or justify their belief in an omnipotent and omniscient God who permitted the brutal murder of more than one million Jewish children under the age of thirteen? Some say that he had "hidden His face," but then, even if we accept the religious interpretation, he could not have ceased being God and therefore must have known what was taking place; in this way, "hiding His face" was a crime of a much more serious nature. Those who purport that we did not follow his path have difficulty contending with the argument that–regardless of the reason, and whether it is understood or not–any justification for murdering children is Satanic. Jewish theologians and thinkers struggle with these questions, to which there may be only one answer: that if the Holocaust was the work of God, then God is not a god but rather Satan.

In light of all this, are we emotionally or intellectually perplexed by the Holocaust? Do we ponder the existential questions that have emerged in its wake? We are certainly perplexed on an emotional level, as the human psyche finds it difficult to accept death in general and mass death in particular. But despite the difficulty involved, there is no way around the issue, and all efforts must be made to accept the Holocaust as historical fact. And if what I wrote above is correct, we must also understand that emotional rebellion against brutal facts have been the lot of mankind since the dawn of humanity. There is no answer – only explanation, the main elements of which I have tried to sketch out above. We must accept the absence of an answer and refrain from trying to seek out the meaning of the Holocaust, because there is none. It has no meaning, with the exception of the meaning for its perpetrators. After all, murder is murder. For the members of the Jewish People, however, contending with the emotional impact of the Holocaust means dealing with it directly – telling the stories of murder, of the flight of the few, of the relatively small number of rescues, and of the heroism of the Jews and non-Jews who saved lives – and, in practice, belatedly engaging in what psychologists call the work of mourning. People instinctively understand this need, which has resulted in a plethora of museums, memorial monuments and plaques, memorial gatherings of survivors from different communities, and personal inquiries into the fate of relatives who died in the Holocaust. It has also resulted in an array of educational efforts to teach about the Holocaust.

Intellectually, the problem is not one of being "perplexed". I am angry, and I attempt to understand. I am not perplexed. The problem, rather, is how to clarify as effectively and accurately as possible what really happened and why. This is also quite difficult, but entirely possible. Methodologically, we must carry on employing two parallel strategies. The first is to engage simultaneously in both micro-history and macro-history. This means asking the big questions regarding motives, general German policy, decision making processes, social structures, economic factors, and other such issues. It also means simultaneously engaging in the examination of small, local contexts, local decision-making processes, the influence of specific individuals, the role of specific groups and organizations such as churches and local authorities and other similar topics, and attempting to integrate the two scales of inquiry. The second strategy is to engage in what we may refer to as the "globalization" of the Holocaust in both a horizontal and vertical manner. By vertical, I mean delving deeply into the historical background of the factors that resulted in the Holocaust specifically, in the context of the factors that have resulted in genocides throughout history generally. By horizontal, I mean studying the context of the Holocaust in world history, European history, and German history in particular; researching the relations between Jews and non-Jews in European countries in the context of the specific developments in those places; studying the history of the

Jews, Jewish communities, and their cultural background; and studying antisemitism, from ancient to modern times, and its impact on the forces listed above (despite the fact that the persecution of Jews was sporadic and that, in most places and at most times, Jews were not persecuted – that is to say, that the history of the Jews is not the history of antisemitism). My assumption that the Holocaust marks a turning point in world history requires a global approach that takes into account what is inaccurately referred to as "Western culture" as a point of reference. Stated in simpler terms, the Holocaust must be studied in its manifold global, historical, social and political contexts, and in any other contexts imaginable.

The most important thing, however, is to never lose sight of the fact that the Holocaust was perpetrated against people, and that personal stories are therefore the most fundamental elements of dealing with the subject. Studying the Holocaust in the absence of an accurate, well-documented story of individuals is a moral and educational failing of the highest degree.

I will conclude with my own personal example. I am engaged in parallel studies. A recent book of mine, *The Death of the Shtetl* (Yale University Press, 2010), examines the small Jewish communities that existed in Kresy, a region which during the interwar period was located in Eastern Poland. My focus was the product of a number of considerations. The most important was the fact that although the murderers have been studied in great detail and much has been discovered about them over the past fifteen years, and although sincere efforts have been invested in assessing the relations between Jews and non-Jews in a number of countries, work on the victims has lagged behind. Despite a number of promising beginnings, work in this direction has been insufficient and has perpetuated images of the Jews as the Germans' object of murder and that of the indifference and hostility of bystanders. This is mistaken. The

Jewish victims of the Holocaust were not objects but rather subjects. They had their own culture, their own societies (in the plural, as the Jewish communities of Europe differed from one another), and their own aspirations. They were the victims of something that was completely beyond their conception, although historically they had been subject to persecution and hatred to a certain degree. But why should they have become a target for extermination? The Jews were never the enemies of the Germans or the German state. On the contrary, the Jews of Germany had done nothing but contribute to German culture. The number of German Jewish men who had fought for Germany during WWI was disproportionately high relative to their percentage of the population. Neither the Jews of Germany, who ultimately accounted for a small minority of the victims of the Holocaust, nor the Jewish victims from elsewhere could have imagined or guessed what was in store for them.

This fundamental premise leads to the conclusion that although the motives of the murderers can be explained (albeit not easily), the reaction of the victims requires thorough and complex clarification. The central issue of every serious confrontation with the Holocaust is the reaction of the victims: How did they respond to the threats that preceded their victimization? How did they respond after they came to understand that the process by which they were turned into victims would conclude in mass murder? An important element of their reaction was their relations with the non-Jewish populations among whom they lived, and the fact that these relationships differed from one another is of great relevance. Polish-Jewish relations differed from Ukrainian-Jewish relations, which differed from Belarussian-Jewish relations, which differed from French-Jewish relations. We know that there is evil in the world, and it fascinates us to the point that we research it *ad infinitum et ad nauseam*. Equally important, however, is how people respond to evil, as every act of mass atrocity has more victims than perpetrators. That is to say, in such a case, the majority of us are most likely to become victims, not attackers. And in actuality, with regard to Darfur, most of us are among the bystanders and in this way also bear guilt.

This was the basis of my effort in *The Death of a Shtetl* to investigate the popular Jewish reaction in the *shtetls* to the evolving situation. It is what underlies my attempt to assess the reaction of the Jews prior to the war, during the Soviet occupation of the region between 1939 and 1941 and during the German offensive, until their deaths. Did the Jews maintain communal unity? In order to raise morale, did they try to maintain the facade of orderly life, mutual assistance, education and child care and basic cultural activities? How did the Jewish leadership echelon (the *Judenrat*) behave under the circumstances? Was there armed or unarmed resistance? Did any traces remain of the pre-war political organizations? Did Jewish culture, which had evolved over the course of thousands of years, continue to exist in various ways under the traumatic circumstances? I know that the answers I provide are lacking, but I hope others will continue the quest.

My research focuses on the study of genocide in general and the role of the Holocaust's place as part of this phenomenon. I am a student of my colleagues who have been pursuing this line of research for some time now, and I try not only to combine what I have learned with what I think I know (perhaps mistakenly), but also to translate it into action within an informal but effective (I hope) framework of international scholars with whom I share subjects of interest. In this way, I attempt to combine academic work with an effort to influence the political world, from the outside of course, as I am not a politician, nor do I strive to be one. However, the idea of being confined to an ivory tower completely repulses me.

Within the same framework, I try to learn and to do my best to influence the struggle against radical Islam, a genocidal ideology that explicitly threatens Jews with another genocide. I do not support the Israeli government's policies toward the Palestinians. Indeed, I am disgusted by them and oppose them with all my might. However, I am also well aware of the threats of genocide against my people and my country. Again, living in an ivory tower is not an option.

Finally, I am engaged in the internal Jewish discussion regarding the nature of Jewish culture and civilization and the role of the Jewish religion. I believe that all these subjects are interconnected: Holocaust, genocide and its possible prevention (or at least its reduction), radical Islam as the most serious and immediate threat of genocide, and the internal Jewish discussion. The Jews, I believe, can be classified as a people whose existence is assured as long as its never-ending internal discussion on the nature of its culture continues. Jews must discuss these issues, as this discussion is the fundamental content of their culture. In this, the Jews are unparallelled. I am happy to be part of this discussion and of the annoying, obnoxious, wonderful and challenging Jewish people. I am glad to have been born Jewish, and if given the opportunity, I would choose to be a Jew.

BOOK REVIEWS

Dariusz Libionka and Laurence Weinbaum, *Bohaterowie, hochsztaplerzy, opisywacze. WokółŻydowskiego Związku Wojskowego* (Heroes, Hucksters and Storytellers: On the Jewish Military Union), Warsaw: Polish Center for Holocaust Research, 2011.

Reviewed by: Jean-Charles Szurek

This study by Dariusz Libionka and Laurence Weinbaum is one of the most innovative books on the Warsaw Ghetto to have appeared in recent years. Focusing as it does on one of the lesser-known and most mysterious topics of the story, namely that of the Żydowski Związek Wojskowy [Jewish Military Union], the Revisionist resistance organization operating alongside the mainstream ŻOB, this work is quite unique. It took an Israeli scholar teamed up with his Polish counterpart to decipher the riddle of this shadowy underground formation. This was so because the source materials are especially meager, and both rumors and even sheer fabrication abound. The authors took it upon themselves to address these difficulties, and have presented us with a two-fold analysis. They describe the entire array of representations of the ŻZW as well as its actual history. It is this two-fold analysis that dictates the structure of the book, with the first section entitled "Deconstruction," and the second, "Reconstruction"—as well as its very provocative title.

The "Deconstruction" section deals with the ideological, political and factual manipulations surrounding the ZW as manifested in Israel, Poland, and elsewhere. The renewed interest in the ZW actually began in 2002 with the publication in France of Marian Apfelbaum's book, *Retour sur le ghetto de Varsovie.*[1] The author of that book is a well-known Paris nutritionist and the son of Emil Apfelbaum, a physician in the Warsaw Ghetto. He sharply attacks the "collective lie" surrounding the history of the Warsaw Ghetto Uprising, which he points out was written and captured by the Polish and Israeli survivors of the ŻOB, foremost among them Marek Edelman. That narrative was supported by the "Yad Vashem historians" all at the expense of the Revisionists, whose military arm in the ghetto was the ŻZW.

Apfelbaum identifies an additional and very significant point of disagreement. He argues that there is strong evidence that Poles fought side by side with ŻZW insurgents in bloody exchanges in Muranowski Square, which was the headquarters of the Revisionist underground movement. According to Apfelbaum, the unfurling of two flags over the ghetto during the fighting—one Jewish the other Polish—definitely demonstrates the solidarity that purportedly existed between Poles and Jews. Despite discontent with Apfelbaum's obviously non-scholarly work, especially with its evident paucity of sources, it did bring the issues to which Libionka and Weinbaum devote their attention into very sharp focus. Following in Apfelbaum's footsteps, Moshe Arens, former Minister of Defense in the Shamir government, joined the effort to confirm this thesis, indicating that the Revisionists played a major role in the uprising. To that end, he published articles in important scholarly journals such as *Yad Vashem Studies, Holocaust and Genocide Studies,* and others.

There is, to be sure, a clear political aspect to this rewriting of history. Libionka and Weinbaum refer readers to the work of Roni Stauber.

This Israeli historian notes that precisely because of the dearth of clear sources on Revisionist fighters in the Warsaw Ghetto, the leaders of the Herut movement (Menachem Begin's party that succeeded the pre-war Revisionists) devoted their attention to glorifying the Irgun (IZL) fighters who resisted the British:[2] "In the early 60s, when Begin and the other leaders of this party came to the realization that their movement had been 'removed from the pantheon,' it was too late."[3] Indeed, the only reference points to the Warsaw Ghetto Uprising were the ŻOB, its history, legend and heroes (Mordechai Anielewicz, Zvia Lubetkin, Yitzhak (Antek) Zukerman, and Marek Edelman)—all associated with the Left.

The ŻZW did not exist in Poland either. The first book devoted to the Warsaw Ghetto Uprising was published in 1946; its author, Joseph Kermish (who immigrated to Israel in 1950 and became the Director of the Yad Vashem Archives) did mention the fighting in Muranowski Square, though without any distinct reference to the ŻZW. Even less attention to the Revisionist underground was paid in the articles published by Warsaw's Jewish Historical Institute (ŻIH), which was charged with chronicling and publishing the history of the Shoah. During the Stalinist period, the uprising was depicted exclusively as an act of the Left, with the Polish Communists portrayed as the main actors.[4]

The memory of the ŻZW, which had been ignored and even erased by various parties, was revived once again in the beginning of the 1960s, thanks to the persistence of several Revisionists who had fought against the Germans during the occupation and who knew about their comrades' role in the Warsaw Ghetto Uprising. Chaim Lazar, a veteran Betari from Vilna and commander of a partisan unit, published a book in Israel in 1963 which became the major work on the ŻZW, *Metzada shel Varsha* [The Masada of Warsaw].[5] Lazar and his wife, journalist Chaya Lazar, traveled to Poland to seek sources for the book and later works, but documentation was scant. One item was a bulletin dated 1943 and attributed to the ŻZW; another was the account by Ringelblum of his visit to the Revisionists' headquarters in the ghetto.[6] There were several other pieces of evidence—but all of them indirect. There was no correspondence or hardly any other written trace of the organization. The ŻZW archives, as noted by Libionka and Weinbaum, were apparently destroyed during the uprising. We do know that the Revisionist institutions existed in Poland until 1949 and that they published ten issues of *Yedioth* in Łodzin 1946-47, but clearly, this data is insufficient to create any chronicle of the ŻZW.

During their visits to Poland, the Lazars met with a number of people, most notably with two Poles, Tadeusz Bednarczyk and Henryk Iwański. Both men claimed to have played a major role in the Polish assistance to the ŻZW. Iwański actually claimed to have been a founder-partner of the ŻZW and to have lost two sons and a brother in the uprising (on April 27, 1943). He also maintained that one Moryc Apfelbaum (to whom Marian Apfelbaum claimed to be related) was supposed to be the commander of the ŻZW, that he himself was his closest friend and that he had witnessed his death in battle that same April 27, 1942. In Lazar's book, Iwański is portrayed as a hero. Lazar spread the word about Iwański's role in the uprising. That information reached Israel and the United States, and of course was disseminated in Poland as well. In 1963, Iwański was awarded the Virtuti Militari Cross, one of Poland's highest medals. In 1964, he became one of the first Poles honored as a Righteous Among the Nations by Yad Vashem, alongside Władysław Bartoszewski and Irena Sendler. Iwański was praised in *The New York Times* and the rescue of thousands of Jews who purportedly passed through his apartment was attributed to him. According to Iwański, the ŻZW was established in 1939 in the first months of the German occupation, operated underground following the Polish model, and had more members than the

ŻOB. Of course, there was no written evidence of any of this information, but according to Iwański, the witnesses were still alive. Chaya Lazar first heard about the existence of Moryc Apfelbaum from Iwański himself.

As a result of the Lazars' lack of familiarity with the reality of Communist Poland and their great distaste for any source "close" to the ŻOB, they fell victim to the trap laid for them by the Iwański-Bednarczykduo and the group surrounding them. An experienced observer might have raised questions regarding the paucity of reliable sources in Apfelbaum's book. Actually, Bednarczyk and Iwański's articles on the Warsaw Ghetto Uprising and the ŻZW mentioned above were in fact published in 1968, i.e., at the height of the antisemitic campaign instigated by the ruling party headed by Władysław Gomułka, assisted by the Minister of the Interior, General Mieczysław Moczar. Even more astonishing is the fact that during the antisemitic campaign, their articles praised the assistance extended by the Poles to the Jews during the uprising.

Libionka and Weinbaum, who had access to the archives of the Polish police (now preserved in the Polish Institute of National Remembrance), succeeded in discovering several new aspects of Bednarczyk's biography. At the time of the Lazars' trip to Poland, Bednarczyk had ties to the S^{3}u¿ba Bezpieczeństwa (SB), the Polish Ministry of Internal Security, and had gained their trust as early as 1962. It is sufficient to read the report by an officer of the Urz^{1}d Bezpieczeństwa (UB), the Polish Secret Police, stating that Bednarczyk and Iwański "did transmit to Chaya Lazar information on the Jewish resistance movement, especially on the ŻZW...which proved the vast assistance provided by the Poles and which emphasized the Jews' lack of determination and lack of desire for resistance against the Germans."[7] In his discussions with the SB, Bednarczyk did not conceal his objective of seeking financial benefit from his meetings with the Israelis, and proposed to the SB that he would write a book to present "the truth to the world," taking into consideration "the needs and suggestions of the Security Service."[8] For Libionka and Weinbaum, "Bednarczyk was very much aware of the growing antisemitic atmosphere that infected the staff of the Ministry of the Interior."[9]

What game was the Polish political police playing in their contacts with representatives of Israel's Herut Party? The truth is that this step was not directed against the Israelis, but against the director of the Jewish Historical Institute, Bernard Mark, with whom they surely had ties. At that time, Mark was under close surveillance, as were many of his contemporaries who were leaders of Jewish organizations. Many of Mark's conversations with foreigners were conducted outside the walls of the Institute, since the building was riddled with hidden microphones.

We could conclude here by saying that this narrative of the history of the ŻZW was actually invented by operatives of the Polish secret police, and that the Revisionists or Revisionist fellow travelers mentioned here (whether the Lazars, Marian Apfelbaum, Moshe Arens, or others) succumbed to it, but that would be too simplistic an explanation. Whoever stops here is evidently unfamiliar with the workings of the Communist system, under which private and relatively independent initiatives by individuals or groups could develop over the years (which, by the way, seems to contradict totalitarian theories). Of course, the interests of such initiatives must never conflict with the strategic interests of the regime. This is certainly the case with the crafty Bednarczyk-Iwański, who sought at a very early stage to make contact with foreign Jews and Jewish institutions outside the Soviet Bloc.

By inventing stories that would please Herut members (for example that Adam Czerniakow and Simcha "Kazik" Rotem took part in the ŻZW), the two, at first separately and then in tandem, tried to appear useful to various representatives of Jewish institutions in Poland and

Israel. Their clear intention was to obtain financial benefit, in one way or another, for their purported assistance to Jewish fighters. Bednarczyk bombarded the Lazars with letters asking for money ("We wrote to you 18 times without any response") and never failed to mention the wounds he sustained in battle on behalf of the Jews in the ghetto. Libionka and Weinbaum's research clearly demonstrates that no members of the Iwański family died during the Warsaw Ghetto Uprising, and that Iwański never even had any sons.

This shameless hoax, disseminated over the course of several decades, is in and of itself a fascinating story. It transcends the history of the Warsaw Ghetto, touching upon the issue of how people become believers, and how such frauds are created and canonized. There is insufficient space here to describe all the details of the extraordinary description of human ties as set forth by Libionka and Weinbaum, but suffice it to mention that the Bednarczyk-Iwański team succeeded in obtaining recognition as Righteous Among the Nations for at least some members of their network, as well as Polish decorations for their comrades. In 1968, the two of them actively supported the anti-Zionist campaign. Iwański died in 1978. His death in Warsaw was not marked by any obituary, and it was only publicized in the Revisionist press in Israel. As for Bednarczyk, after 1989 he joined Polish political organizations on the extreme Right. The two colleagues quarreled at the end, since only one of them received recognition as a Righteous Gentile and the other did not. As a result, Bednarczyk then tried to discredit Iwański and his own role in aiding the Jews.

The history of the Jewish Military Union also has its Jewish storytellers. The authors analyze the works of many apocryphal writers who describe their alleged presence in the Revisionist underground. Such is the case, among others, of Maurice Shainberg and Jack Eisner, who settled in the United States, and published their respective memoirs in the 1980s.[10] These books enjoyed, and in fact still enjoy, some success. Twenty thousand copies of the Polish edition of Eisner's book were printed. Eisner was an influential man in the Jewish community of New York. He lectured widely and was the founder of the Warsaw Ghetto Resistance Organization (WGRO). Libionka and Weinbaum reveal the improbabilities in these books. As for Shainberg, he spent the period of occupation not in the ghetto but in the USSR! The motivations of these scribblers can be understood in terms of both the symbolic and pecuniary satisfaction derived from identification with the heroic history of the insurgents. It is always beneficial to expose the undertakings of those who falsify history, and we should thank the Polish-Israeli historical team of Libionka and Weinbaum for doing so.

After the "Deconstruction" comes the "Reconstruction." If, from the authors' analysis, it turns out that the stories about the ŻZW were mainly political forgeries, how, then, do they face historical reality? Despite the general fog surrounding everything involving the ŻZW, the authors have discovered several new aspects of the story arising from facts on the ground gleaned from new sources.

First, it is clear that the Revisionist movement, especially the Revisionists in Warsaw, suffered from the loss of their senior leadership. At the beginning of the war, headed by Begin, they escaped eastward. Many Zionist youth movements gathered together in Vilna (Dror, Hashomer Hatzair, Betar and others), but for Begin (who remained in Vilna even after the annexation of Lithuania by the USSR on June 15, 1940) and the other Betar leaders, Eretz Yisrael took first priority. In contrast, already by January 1940, Dror activists such as Zivia Lubetkin and Yitzhak Zukerman—who were no less committed to the Zionist cause than their Revisionist rivals—had returned to Warsaw. The absence of Betar's leaders left the Revisionist activists without clear direction, and it took them a long time to re-organize. Libionka and Weinbaum have found nothing to suggest that the Revisionists had begun to prepare armed resistance in 1939, which contradicts the Revisionist historiography. The authors state, "There is barely a single written document mentioning Revisionist activity during the initial months of occupation. Similarly, there is a paucity of documentation on the first period of Betar and Revisionist activity in the Warsaw Ghetto."[11]

In 1941, the Betar movement in Warsaw was indeed rebuilding itself, a fact confirmed by the appearance of a bulletin noting the death of Jabotinsky. In the summer of 1941, a large number of Betar members succeeded in leaving Warsaw for the Hrubieszów area in today's southeast Poland. According to David Wdowinski, one of the extremely few survivors of the ŻZW, this was their attempt to save the Betar youth and prepare them for life in Eretz Yisrael.[12] However, most of these young people were murdered by the Germans in 1942, when they concentrated all of the Jews in the region and deported them to Sobibor.

Between 1942 and 1943, the Revisionist groups that lived in the ghetto founded the ŻZW. According to one of the very few survivors of the Uprising, "Pawel Besztymt" (a pseudonym, real name unknown), who left the ghetto and left written testimony in 1943, the ŻZW was founded in late 1942 or early 1943. He died during the general Polish Uprising in Warsaw in 1944. Both Marek Edelman and Yitzhak Zukerman stated that there were contacts between the ŻZW and the ŻOB. Although it is difficult to describe exactly the nature of these contacts, we do know with certainty that they were severed. Various explanations were offered to explain the break; perhaps the elitism typical of Betar may be the reason.

What is certain is that the ZZW amassed relatively large quantities of arms, which is surprising because we do know that the ŻOB was sorely lacking in weapons. In this respect, the history of the ŻZW as told by Libionka and Weinbaum is unique, since the arms arrived through channels that bypassed the Polish underground movement. The Armia Krajowa [Home Army] (AK) had no idea of the existence of a second military organization inside the ghetto. The members of the ŻZW were undoubtedly young volunteers, but had no wide-ranging contacts with the Polish underground as did the ŻOB.

But what did the ŻZW actually do? Like the ŻOB, they too executed Gestapo collaborators, and also blackmailed wealthy residents of the ghetto to finance their arms purchases. Beginning on April 19, the day the uprising began, they were present at Muranowski Square, but alone there, because the ŻOB was fighting in other areas. Two flags did fly above the square—a blue and white Jewish flag, and the red and white Polish flag, although we have no way of knowing what actually brought about the unfurling of the second one. The symbolic unfurling of the two flags was mentioned by all of the Jewish and the Polish political groups as early as 1944. Prime Minister Stanisław Mikołajczyk referred to it; he considered it a symbol of the brotherhood in arms between Jews and Poles. The Communists made note of it, as did the modern-day Polish Right. As for the Israelis, historians relying on the Bednarczyk-Iwański documents cannot avoid mentioning it either.

For an in-depth understanding of what is now known about the ŻZW, its activities and its young leaders, I refer readers to the meticulous work by these two authors. From their research we learn the identity of the authentic fighters of the ŻZW. Significantly, the story as told by Wdowinski, one of the most important personalities in the Revisionist movement in Poland, and the most senior one present in the ghetto, requires caution. The authors make clear that much of his testimony reveals that he was not an eyewitness to many of the events he described.[13] There are still many unknowns yet to be clarified on all matters involving the Revisionists during the uprising. As Libionka and Weinbaum state in their conclusion, "Major participants in the events perished, and those who remained did not know how or did not want to write this history."[14] The absence or death of the Revisionist leaders led to the growth of new leaders of the organization, Paweł Frenkelfor instance, but the information on him is sketchy and not reliable. The name of Leon Rodal, a veteran Revisionist journalist, also appears. We know that this group of fighters consisted of about 200-250 people, a number which is not insignificant if we remember that the ŻOB represented a broad coalition of several organizations on the Left.

Finally, the history of Betar and the ŻZW in the Warsaw Ghetto—deprived of charismatic leaders, struggling with the loss of its cadres murdered in Hrubieszów—has left no traces. It is a sad story.

We can certainly understand the heirs of the Revisionist Movement, who grasped at straws and sought to weave a heroic legend, despite the facts. In this milestone book, Libionka and Weinbaum have succeeded in providing us with an analysis which separates the wheat from the chaff—examining the entire story and distinguishing mythological constructions from historical truth.

Notes

1 Originally published in French as *Retour sur le ghetto de Varsovie,* by Marian Apfelbaum, (Paris: Odile Jacob, 2002); in Polish as *Dwa Sztandary: Rzecz o powstaniu w getciewarszawskim* (Cracow: Wydawnictwo Literackie, 2003); in Hebrew as *Bekhazara Legeto Varsha: Hitkomemut Ha'igud Hatzvai Hayehudi* (Jerusalem: Yad Vashem, 2004); and in English. *Two Flags: Return to the Warsaw Ghetto* (Jerusalem: Gefen Publishing, 2007).

2 Cf. Roni Stauber, *The Holocaust in Israeli Public Debate in the 1950s : Ideology and Memory*, (London: Valentine Mitchell, 2007), 41
3 Libionka and Weinbaum, *Bohaterowie, hochsztaplerzy, opisywacze,* op.cit.,15.
4 cf. Jean-Charles Szurek, *La Pologne, les Juifs et le communisme* [Poland, the Jews and Communism], (Paris: Michel Houdiard, 2010), particularly the chapter, "Témoin du stalinisme: l'Institut Historique Juif" "Witness to Stalinism: The Jewish Historical Institute".
5 The Hebrew version was published by the Jabotinsky Institute in Tel Aviv in 1963, and the English one, entitled *Muranowska 7: The Warsaw Ghetto Rising,* in 1966. The Hebrew edition was reprinted in 2013 with a foreward by Moshe Arens on the occasion of the 70th anniversary of the revolt.
6 Ringelblum mentioned this visit in his last book, *Stosunkipolsko-żydowskie w czasiedrugiejwojnyswiatowej* [Jewish-Polish relations during World War II], Artur Eisenbach (ed). (Warsaw: Czytelnik 1988).
7 The authors report on this visit of the security officer with Bednarczyk, ibid., 68.
8 Ibid.
9 Ibid.
10 Maurice Shainberg, *Breaking from the KGB : Warsaw Ghetto Fighter, Intelligence Officer, Defector to the West,* New York, 1986 ; Jack Eisner, *The Survivor,* New York, 1980.
11 Ibid., 265.
12 Ibid., 290.
13 Ibid., 41.
14 Ibid., 587.

Woman of Valor Bella Gutterman, *Zivia the One : The Life of Zivia Lubetkin,* Translated from Hebrew by Ora Cummings (Hebrew ed., Hakibbutz Hameuhad and Yad Vashem, 2011), 440pp.; English edition, forthcoming from Yad Vashem Publications)

Reviewed by: Cipora Hurwitz, z"l

I approached the writing of this review with fear and trepidation. Gutterman has devoted very thorough and detailed research to this wonderful biography about a special woman and a unique period in Jewish history. The book is readable, extremely interesting, and written in eloquent Hebrew. Gutterman's biography describes tragic events in Jewish history intertwined with Zivia's persona, supported by many testimonies, quotations from numerous newspapers and the many books written on the subject, periodicals, interviews with Zivia's many friends and acquaintances, her children Yael and Shimon, and from letters left by her husband Yitzhak (Antek) Zuckerman.

Gutterman describes the village of Byteń, Zivia's birthplace, her parents' observant Jewish home and how she left it, yet whose religious directives circumscribed her life. Many young Jews who came of age in the 1920s and 30s distanced themselves from religion, from their parents' way of life and the Jewish leadership to join the Zionist youth movements that blazed a different pathway to their Jewish identity. Zivia Lubetkin was among the young people who joined the *Freiheit* movement against her father's will.

Within a short time, she became prominent due to her intelligence and worldly wisdom. Soon the movement became her home, and the distance from her parents' way of life grew greater. Her devotion to the movement and Zionist ideology engulfed her to the extent that she and the movement merged into a single entity. Although Zivia was a shy and introverted child, when she joined the movement her personality changed: she opened up and had no hesitation in expressing her opinion and thoughts in public. The single personality trait that remained constant was her modesty.

It is difficult to say today how far the extraordinary Zivia would have gone without the events of the Holocaust, which, because of her status in the movement, required her to make fateful decisions at every turn. The events of the Shoah tested her and her comrades in ways that are difficult to describe. These ordeals highlighted her common sense, but also her modesty, moral sense and integrity, despite the shadow of death stalking her at every moment during the Holocaust.

Before the war, the Zionist youth movements founded *hachshara* training farms and collectives called' kibbutzim'. These projects had the young people participate in meaningful activities such as hikes, discussions about the realization of Zionism, sing-alongs and dancing. These activities helped shape Zivia's path and deepened her desire to immigrate to Eretz Israel and establish kibbutzim along the lines of the training kibbutzim in Lutzk and Kielce.

At Lutzk, an emissary from Eretz Israel identified her as a "Hebrew speaker, deeply devout in the pioneering precepts. She was shy, and had an expression of seriousness and maturity beyond her age."

Zivia was sent by the leadership to the agricultural training farm at Kielce, the movement's largest kibbutz, because they believed that she would know how to cope with all the problems the kibbutz was dealing with. Zivia stood up to these difficulties, and it was not long before she became one of the leading figures there. Her powerful strength and decisiveness was already evident, later to be expressed in the Warsaw Ghetto, where she showed her rigor as well as her extreme personal integrity.

In 1938, Zivia was called upon to work in the *Hehalutz* center in Warsaw, and she was appointed coordinator of the movement's training, which integrated her into the movement's national leadership. In August 1939, Zivia was appointed one of the delegates to the 21st Zionist Congress held in Geneva, the first time she ever left Poland. By then, it was absolutely clear to all the delegates that their familiar world was about to collapse. It never occurred to Zivia or her comrades to remain in Switzerland and seek refuge, despite a Certificate having been prepared for her which would enable her to immigrate to Eretz Israel. When the Congress was over, despite the looming threat of war, Zivia returned to Poland to be with the young people in the movement, although the way back to Poland was no longer easy, as many borders had been closed and they had to traverse many countries to return.

By August 1939, after a difficult journey, Zivia succeeded in reaching Warsaw. In Zivia's opinion, which she expressed to Yosef Kaplan of *Hashomer Hatzair*, immigration to Eretz Israel was indeed important, but no less important and pioneering was the decision to remain in Poland and serve the movement.

Immediately upon the outbreak of war on September 1, Kovel and Brisk were bombarded. Knowing this, the next day a central committee meeting of *Hehalutz* in Warsaw was convened, attended by emissaries from other Zionist youth movements to discuss the question,"What should the Jewish Zionist pioneering movement do?" In light of the rapid advance of the German army, the Polish authorities announced that Warsaw would be fortified, with barriers erected in the streets, trenches dug, and all other possible means employed to defend the city.

Upon Zivia's return to Warsaw, which was being bombed and shelled, she did not hesitate to travel to other branches of the movement, using any and all transportation possible, whether by train, cart, or anything available. Throughout all of the discussions held at the *Hehalutz* central committee, not a single person could imagine the bitter end of Polish Jewry; they all spoke of plans for when the war ended, how they would rebuild the training farms and movement branches – it never occurred to them there would be nothing to rebuild! One example was the farm at Grochów, which Zivia considered the symbol of the entire movement, a place which should be maintained at all costs so that after the war it could continue to be a training place for comrades to prepare for *aliya*.

After the defeat of the Polish Army, as the leadership saw reality changing day by day, they decided to head eastward, assuming they would be more protected in the East. Poland, defeated within a few weeks, was split into three parts: central Poland, which became the area under German rule; western Poland, annexed to the German Reich; and eastern Poland, later annexed to the Soviet Union under the Ribbentrop-Molotov Pact. Lithuania remained free until it became a Republic of the USSR. The movement fled eastward as did so many others who wanted to reach the large cities in the east, until they were so crowded that there was no more housing. The youth movement members also strove to reach these big cities.

There is not much known about the period in which the youth movements remained in Eastern Poland and how they operated. Despite high hopes for the Soviet regime, the Soviets harassed the movement in various ways until they were forced to go underground. This was a disappointing experience, but important for the counselors because this complicated, dangerous experience would help the youth movements organize themselves in the underground in the future. Perhaps this helped shape the resilience of the Zionist youth movements, which would later serve them in the ghetto. On September 20, a meeting of the youth movements was held with representatives of all of the Jewish parties. Among the many questions raised was the issue of how to continue to organize the youth, and what methods could be used to continue the immigration to Eretz Israel. Another important question under discussion was how to renew movement activities in German-occupied areas by transferring some of the movement' leaders. In the Soviet sector as well, Zivia became the major factor in the movement leadership despite the deep disappointment in the Soviet regime, and despite the frustration that the Zionist movements had to operate underground, risking imprisonment or disappearance.

Gutterman describes down to the smallest detail what happened to the Zionist youth movements under Soviet rule. The disappointment in the regime in which they had previously believed and the dangers they were facing now became concrete. On December 31, under the guise of the New Year celebrations, about 25 members gathered together pretending to mark the holiday. They debated weighty, ideological issues regarding the USSR, but in my opinion the major issue was the decision that the area under German occupation was their direct responsibility. They determined at the meeting who would remain in the Soviet sector, and who would return to German-controlled Warsaw. Among those who returned to Warsaw was Zivia, in one of her crucial life decisions.

Even though the Soviet area was still safer, and Zivia would be leaving not only comrades but her life partner, this did not prevent her from deciding to return to Warsaw. This is perhaps the place to comment that while the emissaries from Palestine hurried to leave the danger zone, the movement counselors educated in the Diaspora acted impeccably as role models. Zivia's return to Warsaw involved courage, initiative, and stealth so as to avoid Poles.

There were about two million Jews in the German occupation zone. After establishing the General Government and annexing areas to the German Reich, the condition of the Jews greatly deteriorated. For example, it was life-threatening to travel by train for anyone who "looked Jewish," as Jews had to be wary not only of Germans but also of Poles. Those who returned from the East to Poland were only young activists, counselors and leaders who lacked public experience, but who felt deeply responsible for the young members of the movement who remained under German occupation. Their disappointment at the inability to act under the Soviet occupation impacted the decision to return to Warsaw, even though it was controlled by the Germans. They were no more than 20 people – but they were strongly motivated, adhering to the values and principles the movement had inculcated in them.

Immediately upon their return to Warsaw, the counselors began educational activities, bringing the youth together in settings familiar to them from the past, despite the changes dictated by the situation. Among the changes the movement underwent was the thought that they could not confine themselves only to the youth movement, but felt that they were obligated to penetrate to the suffering Jewish masses.

Zivia returned to Warsaw in the winter of 1940 and was amazed and horrified by the huge changes that had taken place in the city in such short a time. In contrast to Polish Warsaw, which had been so bustling and lively, once she came closer to the Jewish Quarter she saw destruction and ruin at every corner. What especially worried her was how the Germans sought out the Jews with beards to rip them off their faces.

As secretary of *Hehalutz*, Zivia did not rest for a moment, constantly busy pleading with the Judenrat or asking for money from the wealthy for the starving masses. She considered this assistance part of her role. Within a short time, Zivia became very well known, and commanded authority and high regard from both adults and young people.

When Yitzhak Zuckerman arrived in Warsaw in April 1940, Zivia began a new chapter in her personal life. Under Zivia's leadership, the *Hehalutz* center in Warsaw became the common supra-framework and the representative of all its members vis-a-vis public institutions in Poland as well as Jewish institutions such as the Joint, ORT, the Judenrat, and others. Her written pleas to Jewish organizations, including letters to Palestine, brought her only disappointment. Her letters screamed out to the free world, "What is the matter with all of you? Where are the rescue efforts? Where are the helpers?" Her disappointment was greater than could be borne. The contact with the outside world, especially with Nathan Schwalb, emissary of *Hehalutz* in Geneva, continued up to the early months of 1942, when communication through letters nearly ceased altogether.

Zivia took no respite from her work. The hunger of the masses, the dead laid out in the streets of Warsaw, the hermetic sealing of the ghetto, all of the aforesaid terrifying problems led Zivia and the youth movement activists to act with increased vigor. Until the ghetto was hermetically sealed, they even made various attempts to establish a farm to provide a minimum of food for the ghetto. However, when the ghetto was sealed, cutting it off from outside economic contacts, unemployment rose, and the number of those working for the Judenrat and the Jewish Police increased. However, despite all this, members of the movement did not give up on establishing seminaries for unemployed youth with prominent lecturers as teachers.

In 1941, when the Germans had already begun genocide in the East, the fact of mass murder was still not internalized by the Jewish public in Warsaw. Not only the masses, but even the Chairman of the Judenrat in Warsaw, Czerniaków, claimed that the reports were exaggerations.

For the heads of the youth movements, receiving information on mass murder in the fall of 1941 was a turning point. They recognized the need to change their *modus operandi* because they had already heard about the massacre at Ponary and were conscious of the fact that "Vilna was the beginning of total destruction." The youth movements understood that they had to seek ways to defend themselves. Zivia and Yitzhak were astounded by the rumors that the murder was methodical, yet they still could not grasp that this was total genocide. The leadership was ready for anything; although they did not think of victory, they thought that if they were going to die, they should die with honor.

When they went to the head of the Joint to consult on how to respond to the Germans, they were cautioned lest they destroy the entire ghetto. Despite the warning, the Zionist youth movements decided to establish a Jewish fighting force. They knew that the basic condition for establishing an underground was the people's support. From the moment of decision to form an underground, the youth movements invested efforts in self-defense on one hand, and in fighting for their honor and the honor of the Jewish People on the other hand. Here, too, Zivia took a major role, mainly in persuading public opinion in the ghetto and its leaders.

It is difficult to describe the ideological struggles that took place at that time under the horrific conditions under which the movements operated, but unifying all of the Jewish organizations was too complicated, despite the situation. The various ideologies still fulfilled a very major function.

The first *Aktion* (roundup) took place on July 22, 1942, when the fighting leadership possessed only two handguns. Already before the *Aktion* the terror had increased in the ghetto, and what may be described as mass murder took place in the ghetto on April 18, which the residents called "The Bloody Sabbath Eve." SS units burst into apartments and shot dozens of Jews. In the ghetto, fear dominated every waking moment, which Gutterman succeeds in describing so vividly that readers feel complete identification. When the large *Aktion* began, and Adam

Czerniakow, as head of the Judenrat, was supposed to hand a list of deportees over to the Germans, he committed suicide. Zivia was very critical towards him, even after his death. She felt that he should have informed the Jews of Warsaw of the Germans' intentions, and had failed to do so. Overall, Zivia was very critical of every Jew who gave in to the Germans' dictates.

On July 23, representatives of the Jewish organizations, excluding the Revisionists, met to decide on a proper response to what was going on. Nevertheless, many were opposed to any resistance activities out of fear, because they still believed that the total liquidation of the ghetto could not be possible. The turning point in how the youth movements viewed events was the big *Aktion*, when they felt that they would be unable to succeed in having all the Jewish organizations agree on armed resistance. They felt that the older people were too cautious and afraid of the outcome. However, the situation was deteriorating, and a rapid decision was needed to found a fighting organization. On July 28, representatives of *Hashomer Hatzair*, *Dror-Hehalutz* and *Akiva* met together and founded the Jewish Fighting Organization (ZOB). Zivia, as was her way, and her partner, were among the representatives of the pioneering movements, and Zivia was the only woman at the headquarters of the ZOB, as a result of her position in solidifying the forces.

Implementing the decision to hold armed resistance necessitated obtaining arms by various means. Because it was dangerous for the men to be on the streets in the Polish side of town, young women who could pass for non-Jews were sent outside the ghetto. From among the men, only those who were fair with light-colored eyes could leave the ghetto. Before the ZOB obtained their first weapon, the Germans had already begun to burn down empty houses from where Jews had been deported to Treblinka, and notices were posted on the walls of inhabited houses.

There is insufficient space to describe all of the ZOB's activities during those fateful days. People need to read the book *Zivia the One,* which contains detailed descriptions of their brave stance under impossible conditions and in small numbers, and how, realizing that their death was imminent, they decided it would not be a passive death in humiliation but a death of courage and heroism, not only for themselves but for all the Jewish People who were being murdered.

The ZOB's battles are described fluently, with many details that enrich our knowledge of what really happened in the Warsaw Ghetto and who Zivia Lubetkin really was.

I have read many books on the Warsaw Ghetto Uprising, but the illuminating details I found in Gutterman's book are unique. The book is very readable, despite the painful details. Zivia's methods and her extraordinary personality are astounding. It was purely by chance that she escaped being killed at Mila 18: when the Germans broke into the bunker with gas, she was out on one of her missions.

Even after immigrating to Eretz Israel with her husband, Yitzhak (Antek) Zuckerman, Zivia did not rest. She was among the founders of her kibbutz, Lohamei Hagetaot, and of the Ghetto Fighters Museum, of which Antek was the director. She was the kibbutz secretary, and as long as her strength held, she stood out with her strong personality.

Kudos to you, Bella Gutterman, for a job well done. I wholeheartedly recommend that readers not miss out on this book.

Without A Trace by Yehuda Bauer, *Death of the Shtetl*, English translation by Atalia Zilber (Jerusalem: Yad Vashem, 2011).

Yehuda Bauer is one of the most important historians of our generation in Holocaust research. His career as an historian began with the study of the period of the struggle for establishing a state (*Diplomacy and Underground in Zionist Policy, 1939-1945*).[1]

Reviewed by: Nili Keren

Through this study, he arrived at the understanding that without addressing the issue of World War II and the Holocaust of the Jewish People, we are unable to understand the struggle of the years 1945-1947 that led to the establishment of the State of Israel. Since his study of the Joint as part of a multidimensional research project which attempted to understand the rescue possibilities and pathways during that period, he has attempted to draw conclusions and find meanings to understand how the Jewish People coped with the unprecedented events of the 30s and 40s.

Despite being a member of Kibbutz Haartzi -Hashomer Hatzair, Bauer's writing and lectures never showed any preference for armed resistance in the ghettos and forests that received particular emphasis by his colleagues who were either members of the same group, or fighters belonging to the movements who returned from Europe after the war. Despite his not having written previously on Jewish history in the pre-Holocaust period, he did not address issues such as "Jewish leadership during the Shoah," but left this to his numerous students at the university, who conducted groundbreaking studies on these and other subjects (Aharon Weiss, for example, on the Judenrat). He always valued the unusually heavy consideration given to armed resistance, and considered this as the most glorious way Jews could confront Germans, but in no way did he consider it the only way.

Bauer was always extremely cautious not to make comparisons between the Holocaust and other cases of genocide pre- and post-Holocaust, nevertheless, he was the one who coined the phrase, "an unprecedented event that became the precedent." Such a statement cannot be so boldly affirmed without having made some sort of comparison between this unprecedented event and events prior or subsequent to it.

In recent years, Bauer has made a very significant turnaround in his academic work, coming back to the study of Jewish life before the war. Not in the large cities bustling with Jewish cultural, political and religious life with its hundreds of thousands of people, nor did he research central and western Europe, where Jews attempted to integrate into society and participate in the civil and national life of the countries in which they lived. Yehuda Bauer returned to the *shtetl,* to the small villages of eastern Poland, Belorussia and the Ukraine. He accompanies the *shtetl* to the execution pits, to their absolute liquidation, their total loss. This is the story of the book under review: *Death of the Shtetl.*

One of the *shtetls* described in the book is Sarny, in the Wohlyn District of Belorussia. This is the village where my mother was born, and where my grandparents were murdered along with three of their children. Nothing remains of them except for a pair of candlesticks my mother received from her mother when she immigrated to Eretz Israel, and a last postcard sent from Sarny on June 15, 1941, about one week before the *Einsatzgruppen,* the SS task forces, entered the area of the USSR, including Sarny.

Everything we knew about the fate of the family came from the memorial book published by immigrants from Sarny and its environs. Relatives who survived the Holocaust in the forests and family camps of partisans knew nothing about the way people went to their deaths.

Immigrants from the *shtetl*, like from other *shtetls*, would meet once or twice a year to tell stories, remember, and mourn those who are no longer with us – but only they, the first generation of survivors, knew the story. The second and third generations grew up without knowing a thing about their families' life and death. A great deal has been written about Warsaw, Kraków, Łodz, and Vilna. Natives of these and other cities wisely established commemorative projects to inculcate the obligation to preserve memory for the second generation. The surviving remnant of the *shtetls* to which Bauer returned were not strong or numerous enough to recreate their story. Now one of the greatest historians of the catastrophe of the Jewish People is attempting to revive the *shtetl*, not only to conduct its funeral or erect its gravestone in the cemetery, but to breathe life into it, if only briefly.

Yehuda Bauer defined the size of a *shtetl* as a Jewish community of between 1,000 – 15,000 Jews. A smaller community would not be capable of founding the abundant number of organizations and institutions that had nothing remotely similar among their non-Jewish neighbors. A larger village would not have enabled close acquaintances between people, or the family atmosphere and human warmth that characterized the Jewish community. The size also limited Bauer's research to a defined district within which he could focus on a number of *shtetls* in eastern Poland, the regions of East Galicia, part of the Ukraine, and Wohlyn.

At the beginning of the book, Bauer devotes space to an historical description of the eras in which he specializes: the 1930s; the German occupation prior to the invasion of the USSR in June 1941; and the period of destruction of the *shtetls*, most of which were liquidated by the end of 1942.

What does Bauer want to teach and learn?

Regarding the 1930s, he concentrates on social and economic developments, the growth of the youth movements, attempts to emigrate, and cultural and ideological developments (23).

On the beginning of the war, he asks the following questions: What was the fate of the refugees? What is known about exiling people to Siberia? What happened to Jewish institutions? How did the Jews get along from the economic point of view? What was the impact of ideology and Communism? (23)

As for the third period: "Here I want to know what the Jews' response was to the unexpected offensive before they were murdered. I know they died. I would like to know how they lived." (23) "I would like to know if the Jewish communities preserved a sense of cohesion under the cruel circumstances before they were destroyed, or perhaps crumbled and were destroyed, and ceased functioning even before their people were murdered? Did they take a stand? (24)

Further on, the book shows that all of the questions Bauer asks are being examined as far as sources permit, and indeed, the question of sources is decisive in research of this nature. The chapter on historical sources is very important, and should be considered not only in the unique context of this book. Every attempt that has been made in the past and is being made in the present to study the life and death of the Jews of Eastern Europe faces tremendous difficulties in finding the documentary material so necessary for the historian to arrive at insights and make generalizations. This chapter makes it clear to us that in the final analysis, the early and later personal testimonies, whether written or taped, are the most important thing, and only as much cross-matching as possible reinforces the description of facts and events. And Bauer is cautious with his statements:

> The postwar testimonies embody another two traps. One – we have the testimonies from a tiny minority, and two –survivors and their testimonies do not necessarily reflect the experiences of the many who were murdered or died. The testimonies were mostly from the middle classes, not the poor, who made up the vast majority of the Jewish community, because most of the people with means succeeded in saving some of their possessions even under the most difficult circumstances, which they could make use of when seeking hiding places in the forests or hiding with peasants who wanted payment for food…the middle class and wealthy survivors are what are reflected in the testimonies we possess. (31)

Asking important, complex questions, the sources used in the research enable readers to address the story of the life and death of the *shtetl* on the eastern frontier, which is the major thrust of the book, with the clear knowledge that perhaps there are more stories somewhere out there, or testimonies that will shake up some of the facts. Consequently, from that point on, Bauer attempts to be cautious in adopting a stance or drawing decisive conclusions regarding the *shtetl* and its residents, their choices, and their behavior at different stages of their lives from the 1930s onward.

I chose to focus on the story of Sarny, the *shtetl* close to my heart and whose history demonstrates the stages of deterioration and death of nearly every *shtetl* described in the book. All of the Jews of Sarny were murdered on August 26, 1942. Up to that day, community life continued, headed by the Judenrat. Bauer wrote thus:

> In Sarny, the more popular Judenrat attempted to aid the local Jews as best they could. There are no testimonies at all of the attempts to provide education for the children, but the religious groups restored activities after they were suppressed by the Soviets. There was a social life and social contacts between people, especially among the younger generation. There was no central handling of orphans and the elderly, but extended families extended assistance. There was no real hunger, and although there was very little food, people used it sparingly, and helped each other. (141)

Later on, Bauer conveys the message that "it seems that there was no underground political activity, but in the last days before the destruction that took place in August 1942, former members of the Zionist youth movements attempted to organize some resistance and failed. (141)

There might not have been fighting in Sarny, but people there did take a social stance: this should not be taken for granted, especially when members of the community took upon themselves to offer material and spiritual mutual aid.

Yehuda Bauer also looks carefully at the "neighbors." As for Sarny, as in other *shtetls*, Ukrainian nationalism and antisemitism were strong, and impacted the local Jews' ability to cope. Nevertheless, there, too, there were demonstrations of assistance and mercy (169, and 177-178). On August 26, all of the 1,634 Jews of Sarny and its environs were concentrated in the square.

> Among the women was a very tall, strong woman, Mindel Eisenberg, called 'Mindel the Cossack' because of her size. Mindel, who was taller than the others, saw a unit of Ukrainians in black uniforms marching towards the fourth side of the square, and immediately sensed what was happening. According to most of the testimonies, she

> screamed out, 'Jews, they're killing us…' A panic broke out, and masses of people attempted to flee…600 to 900 people succeeded in escaping…but at the war's end, there were about 150 survivors, perhaps more. (240)

As we said, Sarny was only one of a number of *shtetls* that constituted only a tiny portion of the many dozens of *shtetls* established in eastern Europe in the former USSR, each with its own story.

The process of decline of the *shtetls* began under the Soviets when they began to repress Jewish life and livelihood. The Germans completed the process of liquidating Jewish life when they murdered them methodically, community after community, Jew after Jew. Yehuda Bauer attempts to revive them from the dead, even if only for a short time, thus enabling readers to get to know a vanished world. The decisive fact is that these *shtetls*, with their men, women and children – our grandparents, aunts and uncles – all lived and live no longer. The book *Death of the Shtetl* is a monument to their memory.

Notes

1 Yehuda Bauer, *Diplomacy and Underground in Zionist Policy, 1939-1945*, (Tel Aviv, 1966) (Hebrew).

The End of the Holocaust, by Alvin H. Rosenfeld, Bloomington, Indiana University Press, 2011.

Reviewed by Guy Miron [Schechter Institue of Jewish Studies. The International Institute for Holocaust Research, Yad Vashem]

How has the memory of the Holocaust been shaped in western society in recent decades and what does the future hold? Is it at all possible to transmit the experience of the Holocaust from the world of the survivors to our own? Where does the border lie between formulation of Holocaust remembrance and the various manipulations that have been perpetrated in its name? These are some of the questions that Alvin Rosenfeld grapples with in this important book.

The basic argument with which the book opens in its introduction is both interesting and enlightening: it is that at a time when interest in the Holocaust became more widespread in the Western world, a change occurred in its basic narrative. Research into the forms of remembrance in recent generations convincingly demonstrates the importance of systems of mediation between events of the historic past, whatever they may be, and the way in which they are perceived and remembered in the present.

With regard to the Holocaust, Rosenfeld maintains that specifically due to the increasing interest in this subject at the present time, an ever-expanding and unbridgeable gap has emerged between the historic event itself ("as it actually happened") and the way the western world perceives "The Holocaust". What motivated Rosenfeld to write this book and focus on these issues was his concern that this growing gap could result in the dwarfing of the perception of this historic and moral catastrophe that the Nazi crimes inflicted upon the Jews, and to a banality of its interpretation.

The impression I came away with from the book was that its nine chapters fall into three parts of unequal length. The first five chapters succinctly lay out the basic theme of the book, as outlined in the Introduction, from different perspectives. The next three chapters describe the works and attitudes of four writers who were extraordinary witnesses to the horrors of the Holocaust and conveyed their experiences to the western world. Rosenfeld demonstrates how - beyond the way these writers present those experiences and their aftermath - they were torn by the tension between their aspiration to pass on Holocaust remembrance to future generations and doubts as to the possibility of ever realizing this aim. The final chapter is an attempt to sum up the diverse aspects discussed in the foregoing chapters.

The Bitburg Controversy – the 1985 visit of US President Ronald Reagan to the German military cemetery where SS personnel were among those buried, and his remark that "these young people too were victims of the Nazis" – is presented in the opening chapter as a significant example of the trend to belittle the memory of the Holocaust and exploit it to promote contemporary causes. While Reagan dutifully visited Bergen Belsen as well, as he was obliged to do, he displayed no interest whatsoever in the true victims of the Holocaust. Behind the US President's rhetoric, which focused on attributing all the evils of the Nazi regime to a single individual, was of course the contemporary aim of promoting German-American reconciliation, and that was out of political motives. According to Rosenfeld this affair, as well as the commercial popularization of Hitler's image – the subject of the rest of the chapter –illustrates the fragile status of the Holocaust in post-war generations, to the present day.

A major factor in the increasing banality with which Holocaust remembrance is perceived is the manner in which it has been exploited in what in Chapter II is described as "the rhetoric of victimization". Rosenfeld argues that the historic suffering of Holocaust victims is today being used almost indiscriminately as a metaphor for suffering of any and every kind. It has been used not only in campaigns for the remembrance of other historic tragedies such as black slavery in America, but also in the context of social causes such as the current campaigns against abortions, for the parental rights of divorced fathers and even as an expression of the frustration of American middle class housewives. The Jewish victims of the Nazi genocide are increasingly caught in the new mathematics of "Victim" in a culture in which suffering becomes the source of identity and of demands for political rights.

This process is just one aspect of the far broader question that is discussed in Chapter III – "the Americanization of the Holocaust". The story of the Holocaust, Rosenfeld argues, has been molded in American mass culture of the past few decades in forms that fit in with the American ethos. Thus it is presented as combining the mystery of the victim with elements such as individualization, demoralization and universal values, while the scale of the evil, the pessimism and the despair are "tamed". The writer views the film "Schindler's List" as an outstanding example of this tendency, which presents the survivors as cultural heroes and transfers the rescuers (in this case Schindler himself) from the margins to the center of the historic event, pushing aside the Nazi crimes that are encapsulated in the persona of Amon Gat, who is portrayed as a psychopath.

The most interesting discussion in the book, in my view, is to be found in Chapters IV and V. It focuses on the process of exploitation and 'taming' of the Holocaust to serve contemporary purposes, specifically relating to the divergent ways in which the image of Anne Frank is presented. What is behind the transformation of this girl and her diary, which has been translated into dozens of languages, into the figure most representative of the Holocaust in so impressive a manner? Rosenfeld discusses the gap between the historic figure of Anne Frank and the contemporary cultural icon that is the teenage Anne Frank, and the gradual process by which this gap came about, viewing this transformation as a test-case for a much broader discussion of the gap between the historic event that is the Holocaust and the view of the "Holocaust" in the contemporary western world.

Whereas the first reviewer of the diary, relating to the manuscript back in 1946 before it was actually published, presented it as an expression of the triumph of beast over man, it gradually, through the fifties, became the bearer of a far more encouraging message. Anne Frank emerged as a naïve, cheerful and optimistic teenager and her story came to be accepted and formulated as a sad but not tragic narrative, and was circulated and received as a kind of drama of adolescence. This elevated reading of the diary emerged as a result of the specific narrative itself being cut off from its historic context. Anne Frank, he argues, has come to be presented not only as a symbol of Jewish destiny but also as an individual who overcame that destiny.

The many readers who looked for an optimistic message in her diary tended to view her prayer to God as a testament of hope, whereas her death in Bergen Belsen, and the testimonies of authentic witnesses who actually saw her in the camp as she was dying – these have been pushed to the sidelines. Rosenfeld shows how, when the Anne Frank story was adapted for the theater, it emerged as an even more sterile, universal and abstract tale, from which Anne's descriptions in the original diary of the unique Jewish suffering inflicted on herself and her family were omitted.

I found of particular interest the analysis of the reception given the German translation of the diary – an extreme interpretation of the persona of Anne that presented her as a universal victim: in this version, the Frank family is detached from its German-Jewish origins, while the term 'the occupation force' is used instead of the term 'the Germans' in the original text. This German translation, according to Rosenfeld, enables the readers to evade confrontation with the truth regarding the sources of the girl's suffering, and strives instead to arouse positive emotions regarding the goodness of mankind. From this point it is a short step to Ronald Reagan's exploitation of the image of Anne Frank in his speech at the Bitburg cemetery for the promotion of US-German reconciliation. At a later stage, this image of the teenager was to become a political icon used in various campaigns, including those against the Vietnam war and for the Palestinian struggle, and has even been used by the North Korean regime in its anti-American propaganda.

In the next three chapters, the discussion focuses on a more concrete aspect of Holocaust remembrance - the way certain survivors grappled with the challenge of bearing witness before the world after the war. Two writers who survived the Holocaust, Jean Améry and Primo Levi, wrote from a sense of urgency, seeking to act as agents of morality after their experience of Auschwitz. But they did not, according to Rosenfeld, elicit an adequate response mainly because what they did was to confront the world with testimony far more horrendous and pessimistic than the 'tamed' version presented in Anne Frank's Diary and in the cinematic treatment in 'Schindler's List'.

Jean Améry, who primarily addressed the German reader, stressed that it was quite impossible to find any valid explanation of the Nazi crimes if one applied the principles of the Enlightenment and its legacy. He committed suicide in 1978, which is viewed by Rosenfeld as a reaction to the ongoing failure of European – mainly German - society to accept responsibility for the crimes of the past. Primo Levi, who tried in his earlier works to project the image of a survivor of great stamina, was also greatly troubled by the question of whether it was at all possible to overcome the public's drive to forget the horrors of the past. In Rosenfeld's view his last work, 'The Drowned and the Saved', reflects his growing fear that the survivors would not be believed. Despite what Levi achieved through his works they lack, according to Rosenfeld, who knew Levi personally in his later years, the element of release.

To offset the awareness of failure that we find in Améry and Primo Levi, in the eighth chapter Rosenfeld discusses two other author-witnesses – Eli Wiesel and Imre Kertesz who, while also tackling recurrent feelings about the ineffectiveness of their works, refused to give in to that trepidation. Rosenfeld analyses various aspects of the works of these two Nobel Prize winners, attempting to pinpoint the elements that protected each of them from succumbing to the feeling of futility that overcame Améry and Levi. Rosenfeld, who is close to Eli Wiesel (in his blurb on the cover of the book he notes that the two have been friends for forty years) maintains that, unlike Améry and Levi, both products of the secular enlightenment who in tackling remembrance of the Holocaust hoped to find understanding and justice in the western sense of these terms, Eli Wiesel focused on the question of the role of the God of Israel in the Holocaust. Unlike Levi and Améry, who were driven to despair by the lack of response to their testimony from the younger generation in Europe, Wiesel's testimony met a far more sympathetic response and wider repercussions in the United States and was never sidelined. For Kertesz, who spent most of his active years in communist Hungary, the limitations imposed by the communist regime served as a kind of protective barrier so that he never developed expectations similar to those of Levi and Améry, and thus was spared the painful and disappointing let-down that they both experienced.

In the ninth and last chapter of "The End of the Holocaust", Rosenfeld attempts to sum up the conclusions drawn from the foregoing chapters and to highlight the current crisis in Holocaust remembrance. This chapter is in my view too eclectic and inconclusive. It is in part a reiteration of the main points of the earlier chapters and allots too much space to the controversy with the critics of the over-manipulative use that is being made of the Holocaust by the State of Israel and the Jewish establishment in contemporary western culture, particularly in the United States. Rosenfeld's criticism of Norman Finkelstein's, "Holocaust Industry" thesis and similar contentions by other writers is not sufficiently analytical in my view, and diverts emphasis away from the problem of the "universalization" and the "banalization" of the Holocaust, which were the main subjects of most of the other chapters – problems that I consider to be far more fundamental than the implications of conspiracy theories such as those advanced by Finkelstein. In this chapter there appears to be no real attempt to connect the conclusions drawn from the riveting descriptions of the surviving witnesses in the earlier chapters with the overall thesis of the book.

Even if one does not unequivocally accept Rosenfeld's approach to the subject of the Jewish character of the Holocaust, and notwithstanding a certain lack of coherence in the overall structure of the book, it nevertheless is an important contribution to the deliberations on the future of Holocaust remembrance in the western world. The very act of formulating the penetrating questions with which the book begins and ends alongside the discussion of various test cases – particularly of the representation of Anne Frank in popular culture and of more complex writings by authors who survived and gave testimony – helps us reach a deeper understanding of world developments. For the Israeli reader, who experiences expressions of Holocaust remembrance in his own environment that differ greatly from current trends overseas, the book is an eye-opener on contemporary cultural developments in the western world.

Translated by Miriam Shimoni

List of Contributors

Prof. Yehuda Bauer, Professor Emeritus. Prof. Bauer is an international authority on Holocaust research. Formerly Head of the Avrham Harman Institute for Contemporary Jewry at the Hebrew University of Jerusalem, and former Head of the International Institute for Holocaust Research, Yad Vashem, and academic advisor of Yad Vashem. He is a member of the Israel Academy of Sciences and an Israel Prize laureate. He was among the founders of the *Moreshet Journal.*

Graciela Ben-Dror, Ph.D, is the Editor of the *Journal for the Study of the Holocaust and Antisemitism* and Former Director of Moreshet. She was a researcher at the Stephen Roth Institute for the Study of Antisemitism and Racism, Tel Aviv University, and a lecturer at the Department of Jewish History at the University of Haifa. Ben-Dror specializes in the Catholic Church during the Holocaust.

Irena Cantorovich, doctoral student at Tel Aviv University. She has researched the history of the Former Soviet Union at the Stephen Roth Institute for the Study of Antisemitism and Racism, Tel Aviv University, and is currently a researcher at the Kantor Center for the Study of Contemporary European Jewry at TAU.

Rivka Elkin, Ph.D., has been researching the activities of Jewish social welfare institutions in Germany under Nazism and has published books and articles on the subject in several academic journals.

Cipora Hurwitz, z"l, a member of Kibbutz Gal-On, survived the Holocaust as a young girl in the Hrubieszow ghetto and the Budzyn and Majdanek death camps. Her memoir *Forbidden Strawberries,* published in Hebrew and English, will soon be published in Polish.

Nili Keren, Ph.D., is a scholar of the Holocaust whose main research interests are teaching about the Holocaust and shaping the consciousness of the Holocaust in Israeli society. She is a veteran teacher at the Kibbutz Seminary College – College of Education, Technology and the Arts, Tel Aviv.

Prof. Matityahu Mintz, Professor Emeritus, Researcher and Lecturer in the Department of Jewish History, Tel Aviv University. He has published several books and studies on Ber Borochov, The Jewish Workers' Movement, and the first steps of the Hashomer Hatzair Movement.

Prof. Guy Miron lectures on Jewish History at the Schechter Institute of Jewish Studies, Jerusalem, and is Head of the Center for the Study of the History of the Jews of Germany during the Holocaust at the International Institute for Holocaust Research at Yad Vashem. Additonal research interests are central Europe in the 20th century; memory, identity and historiography. His most recent book, *The Waning of Emancipation: Jewish History, Memory, and the Rise of Fascism in Germany, France, and Hungary* (2011), was published in Hebrew and English.

Boris Morozov, Ph.D.**,** is a Research Fellow in The Cummings Center for Russian and East European Studies at Tel Aviv University. His main fields of expertise are archival research, Jewish History, Israeli-Soviet relations and the Middle East, and problems of multiculturalism in Israel. Dr. Morozov has published several books and articles, among them his latest book: *Boris Morozov, Soli Shahvar and Gad Gilbar, The Baha'is of Iran, Transcaspia and the Caucasus* (in 2 vols.), 2011.

Dr. Marina Niznik, Ph.D., is currently teaching Russian at the School of Foreign Languages and Russian literature at The Cummings Center for Russian and East European Studies at Tel Aviv University. She is involved in several projects that explore the acculturation of Russian immigrants in Israel. Dr. Niznik is the author of several Russian textbooks for foreigners as well as numerous articles.

Dov Schidorsky, Ph.D., taught at the Hebrew University School of Library and Archive Studies and served as its Director from 1978-1989. He was one of the first employees of the IDF Archives and the State Archives, and one of the founders of the Tel Aviv University Library.

Jean-Charles Szurek, Ph.D., is Emeritus research director in the CNRS at the Institut des Sciences Sociales du Politique (CNRS-Université Paris Ouest-ENS Cachan). His scientific activities are devoted to two fields: 1. The communist world and post-communist transition, particularly in Poland, 2. History and memory of Jewish-Polish relations since World War II.

INSTRUCTIONS FOR CONTRIBUTORS

MORESHET, THE JOURNAL FOR THE STUDY OF THE HOLOCAUST AND ANTISEMITISM

1. Moreshet Journal invites researchers to send us scholarly articles and/or review essays relating to topics dealing with the Holocaust and Antisemitism.

2. Articles have to be original, and not published in other publications.

3. Manuscripts should not exceed 12,000 words including notes, and including a short abstract of 150 words. The articles must be sent by email or on CD, must be typewritten, double-spaced, and all pages should be numbered consecutively. Two typewritten, double-spaced copies of the article on A4 paper should be sent to the address of the Moreshet Journal.

4. Details of the author's institutional affiliation, degree, short CV for the list of contributors (3-4 lines), address, phone number and email must be added.

5. To be accepted for publication the articles will pass both internal and external peer review and, if accepted, the author may be requested to make revisions.

6. Articles may be sent in English, French, Polish or Hebrew, but in some cases other languages will also be considered.

7. The author will receive 10 offprints of the article and a copy of the volume in which the article was published.

Please send the articles to the following address and e-mail:

Dr. Graciela Ben-Dror, Editor
Moreshet, the Journal for the Study of the Holocaust and Antisemitism
Givat Haviva, Doar Na Menashe, 37850
Israel
graciela.moreshet@gmail.com

Made in the USA
Monee, IL
08 November 2026

67994714R00144